Coffee - Production and Research

Edited by Dalyse Toledo Castanheira

Published in London, United Kingdom

IntechOpen

Supporting open minds since 2005

Coffee - Production and Research
http://dx.doi.org/10.5772/intechopen.82913
Edited by Dalyse Toledo Castanheira

Contributors
Cezar Francisco Araujo-Junior., Vinicius Cesar Sambatti, João Henrique Vieira Almeida Junior, Henrique Hiroki Yamada, Saithong Phommavong, Maliphone Douangphachanh, Khanhpaseuth Svengsucksa, Bruno Montoani Silva, Geraldo Oliveira, Erika Silva, Milson Evaldo Serafim, Carla Eloize Carducci, Samara Martins Barbosa, Laura Melo, Walbert Junior Santos, Thiago Reis, César Oliveira, Paulo Guimarães, Julieta Andrea Silva De Almeida, Won-Hee Kang, Mesfin Haile, Laura Sofia Torres Valenzuela, Johanna Andrea Serna Jimenez, Katherine Martínez Cortínez, Hemraj Sharma, Felipe De Jesús Cerino Córdova, Azucena Garcia, Nancy Dávila, Jacob Salazar, Eduardo Soto-Regalado

First published in London, United Kingdom, 2020 by IntechOpen
IntechOpen is the global imprint of INTECHOPEN LIMITED, registered in England and Wales, registration number: 11086078, 7th floor, 10 Lower Thames Street, London, EC3R 6AF, United Kingdom
Printed in Croatia

British Library Cataloguing-in-Publication Data
A catalogue record for this book is available from the British Library

Additional hard and PDF copies can be obtained from orders@intechopen.com

Coffee - Production and Research
Edited by Dalyse Toledo Castanheira
p. cm.
Print ISBN 978-1-83880-884-6
Online ISBN 978-1-83880-885-3
eBook (PDF) ISBN 978-1-83880-886-0

For EU product safety concerns:
IN TECH d.o.o., Prolaz Marije Krucifikse Kozulić 3, 51000 Rijeka, Croatia, info@intechopen.com or visit our website at intechopen.com.

Meet the editor

Dalyse Toledo Castanheira has a BSc (2013), MSc (2015), and PhD in Agronomy (2018) from the Federal University of Lavras (UFLA). She also completed a postdoctoral program at UFLA with a scholarship from the Coffee Research Consortium. She has experience with coffee crops, with emphasis on their sustainable management. Dr. Castanheira has worked on the study of agronomic techniques to optimize the use of water in order to mitigate the effects of climatic variations on coffee culture. She is currently professor and researcher at the Federal University of Viçosa (UFV). She works in teaching, research, and technology transfer activities (with direct relationship with coffee growers) focusing on the sustainability of the coffee production system.

Contents

Preface

Coffee - Production and Research presents a diversity of important issues related to coffee, with an emphasis on the science of coffee growing. Coffee is one of the highest value commodities traded worldwide. Cultivated and consumed widely, it generates progress for both the economy and society.

Coffee growing is located in tropical and subtropical regions. Coffee can be exported and imported in different types: green beans, roasted and ground coffee, and soluble coffee. Production is significant in about 60 countries from different continents. Coffee production stands out on the American continent, followed by Asia and Africa (origin of coffee).

There are more than a hundred coffee species in the world, but only two have economic value: *Coffea arabica* L. and *Coffea canephora* Pierre ex. A. Froehner. The other species are important too, but in general they are more used in breeding programs.

On a large scale, there are countries that grow only Arabica coffee (e.g., Colombia), some that grow only canephora coffee (e.g., Vietnam), and a few that grow both (e.g., Brazil).

Currently, there is a growing need to generate knowledge and technologies capable of optimizing the production of coffee in different types of cultivation, mainly to obtain a more productive and sustainable coffee crop in the face of current challenges.

Chapters in this book cover such topics as biotechnology, growing, harvesting, post-harvest handling, quality, chemistry, commercialization, and byproducts of coffee.

I am very grateful for the opportunity to edit this book. I want to thank the authors and co-authors who contributed to the construction and quality of this volume.

Dalyse Toledo Castanheira
Professor of the Agronomy,
Department of the Federal University of Viçosa,
Brazil

Section 1

Biotechnology

Chapter 1

Observations on Somatic Embryogenesis in *Coffea arabica* L.

Julieta Andrea Silva de Almeida

Abstract

Somatic embryogenesis contributes to coffee breeding programs. This is a process of asexual reproduction which is based on the concept of cellular totipotency. Each haploid or somatic cell of the plant tissue has the genetic information necessary to generate a complete and functional plant. The somatic embryogenesis can occur either indirectly or directly. *Coffea arabica* genotypes may respond to direct, indirect, or both. In this species, the indirect somatic embryogenesis is composed of two phases, the callogenesis and the embryogenesis, while the direct pathway occurs in a single phase, without the callogenesis. In *Coffea*, in general, the indirect pathway is induced by the auxin and cytokinin, and the direct pathway with cytokinin only. *C. arabica* genotypes usually respond easily to the indirect route with high production of somatic embryos. But these are inefficient by the direct route because they present low production of embryos and the process occurs for a long time. In this review, emphasis will be given to different events that are part of the somatic embryogenesis of *C. arabica* occurring indirectly and in the direct pathway as well as factors that may affect its control.

Keywords: Callus, leaf explant, plant hormone, embryogenic structures, somatic embryos

1. Introduction

The genus *Coffea* belongs to the Rubiaceae family and has over 100 species [1]. However, only *Coffea arabica* and *Coffea canephora* species are commercial, the former accounting for almost 75% of world coffee production and the latter for the remaining 25%. Brazil is the world's largest producer and exporter of coffee accounting for about 70% of world exports [2]. *C. arabica* has this hegemony for producing pleasant and stimulating drink that is consumed worldwide while *C. canephora* coffee is less palatable and is intended primarily for the instant coffee industry.

Coffea breeding aims to combine genotypes of *C. arabica* and *C. canephora* species to release genetically stable varieties with strong traits of both species [3, 4]. In conventional *Coffea* breeding it takes six to eight selection cycles to generate a new cultivar, which is about 40 years. Each cycle corresponds to five years. But it takes four to five harvests to consistently evaluate a generation [5]. In addition, it is important that evaluations are performed on plants over five years of age to obtain reliable yield data [6]. In the breeding program, during the selection phase intermediate populations of progenies are generated and each one of them has different genetic pattern. Thus each progeny corresponds to a single plant.

In the selection phase, the progenies have the characteristic of heterosis that favors the occurrence of differentiated plants. Some of these plants may have special characteristics such as tolerance to biotic and abiotic factors or high productivity or excellent drink quality or all of these. To confirm if a progeny is special it must be multiplied and evaluated in relation to its agronomic performance in the field. Following this phase, the progeny may be released as a clonal cultivar. Cloning selected materials allows to capture all selection and improvement gains without involving genetic segregation [7].

The multiplication of intermediate genotypes to breeding program is not indicated by seeds because plants resulting from the germination may have genetic segregation that leads to loss of the special features [8]. Thus, it is recommended that these genotypes be vegetatively multiplied to maintain their genetic pattern. The vegetative multiplication of coffee plants has been obtained by cutting, and the species *C. canephora* responds very well to this process [9] while arabica plants are less efficient by this way, having low multiplication rate [10].

Usually *C. arabica* genotypes are vegetatively multiplied by in *vitro* cultivation. *In vitro* culture or plant tissue culture belongs to Plant Biotechnology which comprises culturing cells, tissues or organs under aseptic conditions and using artificial culture media containing different components such as water, minerals, vitamins, carbon source and plant growth regulators [11]. Plant tissue culture involves micropropagation processes such as somatic embryogenesis.

2. Somatic embryogenesis

Somatic embryogenesis is defined as a process in which a zygotic embryo-like bipolar structure develops from a nonzygotic cell with no vascular connection to the original tissue [8, 11–14]. This process is based on the concept of cellular totipotency, where all the somatic cells of plant tissue contain the genetic information necessary to produce a complete and functional plant (Haberlandt 1902 *apud* [15–17]). In somatic embryogenesis there is no gamete fusion, only somatic cells of explant tissue that will be responsible for the formation of somatic embryos. These embryos undergo the same developmental stages as the zygotic embryo (**Figure 1**) that will develop into a plant with a genetic pattern identical to the explant donor plant [18].

The occurrence of somatic embryogenesis is associated with the induction of differentiated explant tissue cells that acquire the embryogenic characteristic, followed by the expression of the somatic embryo [19–22]. Thus, this process consists in the termination of the gene expression model present in the explant differentiated tissue cells that will be replaced by the expression of embryogenic genes [23, 24]. But embryogenic program does not occur in all cells at the same time, only in some of them [14, 17, 25, 26]. Several changes may occur to reprogram a somatic cell to the competent embryogenic stage.

Embryogenic cells have different characteristics, they are unique, small, superficially they are similar to meristematic cells, with isodiametric forms, small vacuoles, stained nuclei, with abundance of organelles, thick cell wall and starch accumulation [22, 26–28]. The formation of somatic embryos is strongly associated with the embryogenic competence of the explant cells. Possibly, the acquisition of embryogenic competence is related to the endogenous level of plant hormones, which favor tissue sensitivity to plant growth regulators present in the culture medium, which modulates events leading to the formation of the somatic embryo [21, 28, 29].

Somatic embryogenesis can be applied to most plant species [30], but adequate conditions must be available for this, such as explant type, culture medium and growing environment condition [28].

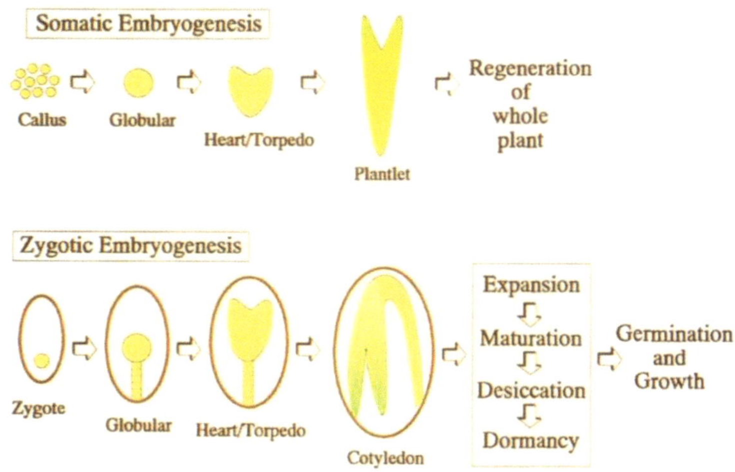

Figure 1.
Schematic representation of the comparison of somatic and zygotic embryogenesis from Ref. [18].

3. Somatic embryogenesis in *Coffea*

Somatic embryogenesis is applied to *C. arabica* and *C. canephora* species for the purpose of vegetative multiplication of elite cultivars, large-scale clonal cultivar, Arabica F1 hybrid [4], to obtain transgenic plants and also as plant differentiation study model [31]. In addition, cloning allows the *in vitro* stock of germplasm cultivars for exchange between research institutions and the preservation of materials since seeds of this species have some degree of recalcitrance [32, 33].

In *C. arabica* the formation of somatic embryos can be obtained either by indirect [34], direct [35] or both somatic embryogenesis (**Figure 2**). Indirect somatic embryogenesis occurs in two phases, callogenesis followed by embryogenesis. On the other hand, direct somatic embryogenesis occurs in one phase without callogenesis. Comparison between these two forms of embryogenesis shows that the direct pathway is more advantageous than the indirect pathway. The direct pathway that occurs in a single phase ends up reducing inputs, labor and cultivation time while the indirect pathway occurs in two phases with higher costs of producing somatic embryos [6, 36, 37]. But although the direct pathway is more advantageous, most studies show that vegetative multiplication of *C. arabica* genotypes is mainly achieved by indirect somatic embryogenesis [6]. In the indirect pathway, explants of this species produce embryos more easily and in high quantity while in the direct pathway the number of formed embryos is smaller and this process occurs in a long time. Studies have indicated that the difficulty of direct pathway occurrence in *C. arabica* genotypes seems to be related to the presence of substances produced by the explant tissue itself [38]. This was demonstrated in explants of *C. canephora* and *Daucos carota* that showed inhibition of direct route when they were cultivated in culture medium with addition of substances secreted by explants of *C. arabica*. This result is possibly related to the difficulty of direct pathway occurrence in *C. arabica* genotypes.

Coffee regeneration via somatic embryogenesis can be obtained from different types of explants (**Figure 3**), such as anthers [39, 40], leaves [34, 41–46] and roots [47, 48]. However, leaf explant is the most used type for the application of direct or indirect somatic embryogenesis in *C. arabica* genotypes, this has been occurring since the pioneering work of Sondhal and Sharp [34]. Normally, these explants are used in rectangular shape with dimensions close to 1.5×2 cm^2.

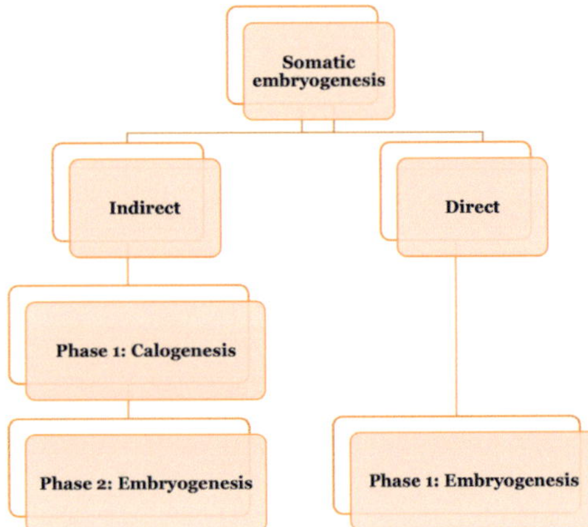

Figure 2.
Details of the occurrence of indirect and direct somatic embryogenesis in Coffea.

Figure 3.
Types of explants used for the application of indirect and direct somatic embryogenesis in Coffea *genotypes.*

There are indications that *Coffea* somatic embryogenesis is more efficient if applied to leaf explants from *in vitro* seedlings than from leaves collected from plants in the natural environment [49–51], and this response must be related to vegetable hormones endogenous. This aspect may also be related to morphology since *in vitro* seedling leaves have a thicker cuticular layer than those from environmental plants. This characteristic tends to favor greater absorption of culture media components leading to the efficiency of somatic embryogenesis response [51, 52]. In addition, *Coffea* leaf explants from environmental plants can become curled which impairs nutrient absorption from the culture medium.

C. arabica explants remain green until about 60 days after inoculation in the culture medium and after this period they oxidize. This response is verified in explants submitted to direct or indirect route [45, 46]. In general, oxidized explants also have the ability to form calluses or embryos although they appear to be senescent.

In the indirect pathway, somatic embryos of *C. arabica* are formed after about 210 days of explant inoculation in the culture medium [45] while in the direct pathway they can be observed after about 90 days, but in small numbers [46]. On the other hand, there are also genotypes that do not form somatic embryos when submitted to these two embryogenesis pathways and are called recalcitrant [53].

Leaf explants of *C. arabica* cultivar Mundo Novo form somatic embryos in both pathways [45, 46]. But the same genotype may have different capacity for somatic embryogenesis in each of these pathways. Explants of cultivar Mundo Novo and decaffeinated genotypes AC1, AC2 and AC3 formed a greater number of somatic embryos via the indirect route than by the direct one [54].

3.1 Indirect somatic embryogenesis in *Coffea*

In *Coffea*, indirect somatic embryogenesis occurs in two phases, the first is callogenesis followed by embryogenesis that corresponds to the formation of somatic embryos [34, 45, 55–57] (**Figure 4**). Another characteristic of the indirect somatic embryogenesis in this species is the occurrence of somaclonal variation in cloned plants, due to the long time that callus remains *in vitro* [58]. Somaclonal variation is undesirable because it leads to the formation of mutants, which can compromise plant growth and development. On the other hand, its occurrence is desirable to obtain genotypes with genetic variability that can be incorporated into the coffee breeding program [59, 60]. For the induction of explant mutations are cultivated at high concentrations of 2,4 D.

Most studies use the Sondhal and Sharp protocol [34] for the application of indirect somatic embryogenesis in *C. arabica* genotypes and these usually respond with some somatic embryo production even when they have low regenerative capacity.

3.1.1 Callogenesis

For the induction of callogenesis in *Coffea* genotypes in general it is used the protocol with MS medium [61] and the addition of phytoregulators 2,4 D and kinetin and 30 g/L sucrose [34]. At this stage, auxin 2,4 D is used at high concentration [26, 62–65] which causes disturbance of endogenous auxin metabolism of explant tissue leading to cell division [8]. Auxin stress is required to obtain calogenesis in most species [12, 66]. Thus, in the induction of callogenesis, differentiated somatic cells of the explant tissue undergo re-determination, with the occurrence of cell division and proliferation events that form a non-functional cell mass, the callus. But it is also found that the induction of calogenesis in *Coffea* can be obtained in response to the use of other auxin types such as NAA [67, 68] and Picloram [69, 70].

Sucrose is used in high concentration to provide energy for the induction of calogenesis [34]. Biochemical analyzes in *C. arabica* explants indicated the occurrence

Figure 4.
Indirect somatic embryogenesis in the cultivar Mundo Novo of Coffea Arabica*. (A) Leaf explant with callus presence 15 days after the beginning of cultivation. (B) Calogenesis at 40 days of cultivation. (C) Calogenesis at 90 days of cultivation. (D) Callus with somatic embryos. (E) Embryonic axis. (F) Seedlings. (G) Plant transferred to* ex vitro *environment.*

of high concentration of soluble sugars in the calogenesis phase while it was lower in the formation and maturation of somatic embryos [71].

In the indirect pathway, *C. arabica* leaf explants initiate callus formation from procambium cells around the seventh day after the beginning of cultivation [72]. In general, from the 14th day of cultivation, it is possible to see the occurrence of small callus on the edges of explants that reach up to 3 mm in size [45]. These calluses are usually formed on only one or two sides of the explant and later reach sizes that can occupy the four edges of the explant. In early development, calluses are hyaline, but gradually they tend to oxidize. In a set of explants of the same origin subjected to indirect pathways, it is possible to find calluses that do not develop. They remain small, up to 5 mm in size. However, some of these calluses have the capacity to form somatic embryos, which indicates that they appear to have been "latent" during this time. Furthermore, it is further noted that not all *C. arabica* explants form calli indicating that the somatic embryogenic capacity may vary between explants from the same plant or from the same leaf.

Callus of *C. arabica* can reach sizes up to 30 mm around 90–120 days from the beginning of cultivation. During this period, these calluses may also become oxidized. Oxidation of explants and callus is associated with the high content of phenols present in tissues of this species. It is also observed that normally oxidized calli have the capacity to form somatic embryos although they have an appearance of senescence.

3.1.2 Embryogenesis, embryo induction

For the second phase of the indirect route, MS/2 medium is used, with NAA and kinetin added and 20 g/L sucrose, according to the protocol of Sondhal and Sharp [34]. For this purpose, calli are transferred from the calogenesis induction medium to the embryogenesis medium. In this phase, there is the initiation and development of somatic embryos from certain cells, located in some sectors of the callus, which correspond to embryogenic centers [45, 55, 57]. Somatic embryos are usually formed on the bottom of the callus, which is in contact with the culture medium and also on the top surface of the callus. On the other hand, not all callus form somatic embryos.

The induction of embryogenic cells occurs during a precise moment of callus life [19]. This induction window may vary for identical explants of the same genotype depending on the cultivation conditions used and particularly the hormonal balance [20, 73]. In the same explant, it is possible to find competent or non-competent callus sections, which indicates that genetically identical cells may respond differently to and from a particular stimulus. Embryogenic cells rapidly lose their ability to divide due to the occurrence of the differentiation event, so the moment of callus transfer to embryogenesis is crucial. As the callus gets older, embryogenic cells lose their specificity in forming embryos. Somatic embryo formation is a continuous process and several embryonic stages can occur at the same time, in the same explant, in the same culture [74].

3.2 Direct somatic embryogenesis in *Coffea*

The process of direct somatic embryogenesis in *C. arabica* occurs in a single phase, which is the main feature of this pathway [75] (**Figure 5**). In this case, explant tissue cells are already determined and competent [76] for embryogenic development, before being extracted from the explant donor leaf [77] and may differentiate into somatic embryos soon after the start of cultivation. In this pathway, embryo formation occurs from leaf explant mesophyll cells [27]. The phenomenon of direct formation of embryogenic tissue in the explant is described as cloning of certain cells [78]. This cloning is a kind of large-scale reproduction of pro-embryos. Thus, pre-embryogenic cells present in explant tissues are cloned and require less epigenetic reprogramming compared to determined embryogenic cells in the indirect process. In the direct route, the need for cellular reprogramming is different from the indirect route.

To respond to the direct pathway, *Coffea* explant cells only need contact with the cytokine-like plant growth regulator and the auxin normally inhibits its occurrence [79]. Explant edge cells differentiate into somatic embryos in response to cytokine in the culture medium [62, 80–83]. The efficiency of the direct pathway also seems to be related to explants from young leaves. On the other hand, in direct somatic embryogenesis, the occurrence of somaclonal variation tends to be minor or absent, since somatic embryos are formed in a single phase.

In the direct pathway, after inoculation of the explants in the culture medium, the formation of embryogenic structures occurs, which can be visualized around the 15th day of culture [75] (**Figure 5A**). Embryogenic structures are formed,

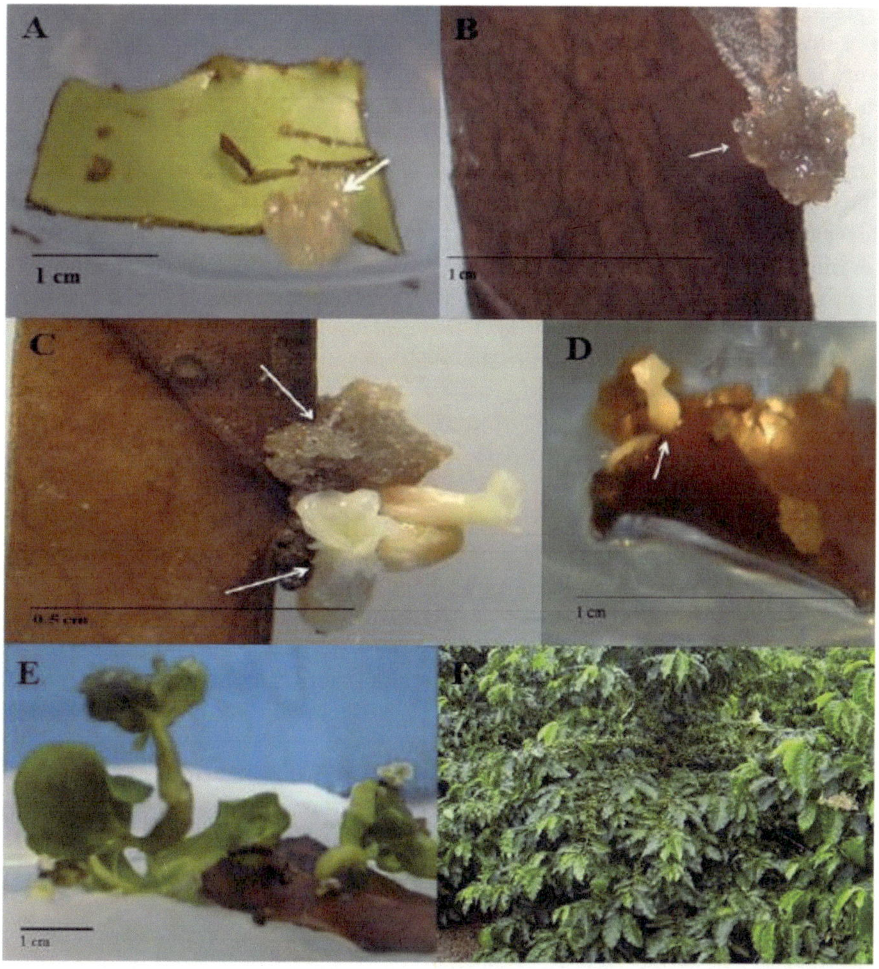

Figure 5.
Direct somatic embryogenesis on explants of C. arabica *Catuaí cultivar maintained in the dark, at 30°C, for 320 days from the beginning of the experiment. (A) Embryogenic structure (arrow), up to 15 days; (B) oxidized embryogenic structure (arrow), up to 70 days; (C) embryos (lower arrow) formed from the embryogenic structure (top arrow), up to 100 days; (D) embryos formed from the border of the explant (arrow), up to 100 days; (E) Plantlets at 320 days; (F) adult plants [75].*

on average, on one or two sides of the rectangular leaf explant. Normally, these structures range in size from 2 to 4 mm (**Figure 5B**) and remain this way until the end of cultivation. About 50 days after the start of cultivation the structures start to oxidize (**Figure 5C**) and by 150 they are completely oxidized [75]. The formation of somatic embryos usually starts from 90 days of cultivation but in low quantity and around 120 days this number tends to increase. Somatic embryos in addition to being formed at the edges of explants (**Figure 5D**) also develop over the surface of embryogenic structures (**Figure 5C**).

3.3 Somatic embryo of *Coffea*

The main product of somatic embryogenesis is the somatic embryo. Somatic embryos are structures that go through different stages of development until they reach that of a plant [18, 28]. These stages of development are perfectly organized, with all the morphological characteristics corresponding to the same stages of development of zygotic embryos [18, 84, 85] being globular, heart, and seedling. Somatic embryos are bipolar morphological structures, presenting radicle, hypocotyl and cotyledons. The somatic embryo does not exhibit endosperm differentiation and is independent of explant tissue after initiation and development. Somatic embryos do not go through the maturation or desiccation phase as in zygotic embryos. This system is not connected to the vascular tissue of the mother or explant during its initiation and development [86]. Moreover, in the direct pathway, it is possible to find somatic embryos at different developmental stages in the same explant [16]. This response pattern can also be found in the indirect callus.

Somatic embryos formed by the direct and indirect pathways are transferred to the germination stage and generally use MS/2 culture medium added 20 g/L sucrose and without plant growth regulators. This same medium can also be used for embryo growth and development to the seedling stage. These observations suggest that *Coffea* somatic embryos may have a hormonal balance that favors differentiation of developmental stages, requiring only nutrients from the culture medium without phytoregulators for germination and growth and development.

4. Factors that influence the occurrence of somatic embryogenesis

The control of the occurrence of somatic embryogenesis in *Coffea* is not yet completely identified. Some authors relate the genetic pattern of the species with the absence or low responsiveness [87]. Knowing the factors that control the occurrence of somatic embryogenesis in *C. arabica* will allow to optimize its application and especially the direct pathway. The high or low capacity of somatic embryogenesis of a species is related to the presence of competent cells or not in the explant, inherent to their totipotency [88]. The maintenance of somatic embryogenesis capacity requires the use of conditions that maintain the proliferation of determined and competent cells [86].

Somatic embryogenesis regeneration capacity is also associated with other factors such as explant donor plant developmental stage, explant donor plant physiological conditions, explant position relative to the plant [89], *in vitro* culture conditions and mainly of plant growth regulators. The seasons influenced the indirect somatic embryogenesis response of plant explants to eight *C. arabica* genotypes in the field [56]. Explants formed more somatic embryos in the fall-winter season than in the spring-summer season.

4.1 In vitro culture condition factors

4.1.1 Lighting

In indirect somatic embryogenesis of *C. arabica*, calli are induced and initiated in the absence or presence of light, but they reach a larger size only if maintained in the absence of light [45, 86]. The size of these callus increases gradually each month, and can reach sizes up to 30 mm.

On the direct pathway, *C. arabica* explants also have difficulty responding in the presence of light. In this way, at the edge of the explants, small structures are formed, which are called embryogenic structures, which remain without change in size and shape in the presence or absence of light.

4.1.2 Cultivation temperature

Cultivation temperature is another factor that may influence the somatic embryogenesis response. Leaf explants of cultivar Catuaí and two hybrids showed higher formation of somatic embryos at 30°C compared to 25°C [90].

4.2 Plant growth regulation

Several studies indicate that phytoregulators play a decisive role in controlling the formation of somatic embryos in *Coffea* leaf explants, which is the most explored aspect on this subject. For the induction of the indirect pathway in *C. arabica*, the auxin/cytokine combination, which is already well established for this species, is generally used. In this pathway, auxin 2,4 D has been the most used to induce callogenesis in *C. arabica* leaf explants. This auxin is considered strong and is also used for the induction of anthers [39, 40] and roots [47, 48].

For induction of the direct pathway, most studies use cytokine without auxin because it tends to inhibit its occurrence. However, the efficiency of the direct pathway response may vary depending on the type and concentration of cytokine employed. A pioneer study showed that 6-BA at 5 μM dose matched direct pathway induction in *C. arabica* explants [91].

The formation of somatic embryos was also obtained from *C. canephora* explants inoculated in MS medium only with addition of different cytokines being 2-iP, ZE, Ki and 6-BA, all at a concentration of 5 μM [92]. Explants formed somatic embryos in the presence of all cytokines, but responses varied according to cytokine type. The 2-iP was more efficient than the ZE, Kin and 6-BA. In this study, it was also found that auxins used at different concentrations inhibited the direct somatic embryogenesis of these genotypes. In another study, Zeatin caused the direct pathway response in *C. canephora* explants [87]. Cytokine 2-iP also caused direct pathway induction in *Coffea* [93, 94]. In another study, it was found that 2-iP concentrations of 7.5 and 10 μM favored a greater number of somatic embryos than the 2.5 and 5 μM doses [95].

Synthetic cytokine 6-benzylaminopurine also has the ability to induce the direct pathway in *C. arabica* explants [69, 91, 96–98]. The 6-BA used at 30 μM concentration caused higher production of somatic embryos than the 10 and 20 μM doses in leaf explants of the cultivar Mundo Novo de *C. arabica* [46]. But although the 6-BA concentration was high, embryo production was reduced and the process also took place over a long time. However, this result is interesting because it favors the cost reduction of clonal seedling formation since 6-BA is a cheaper and available synthetic cytokine than zeatin and 2-iP. Cytokine TDZ

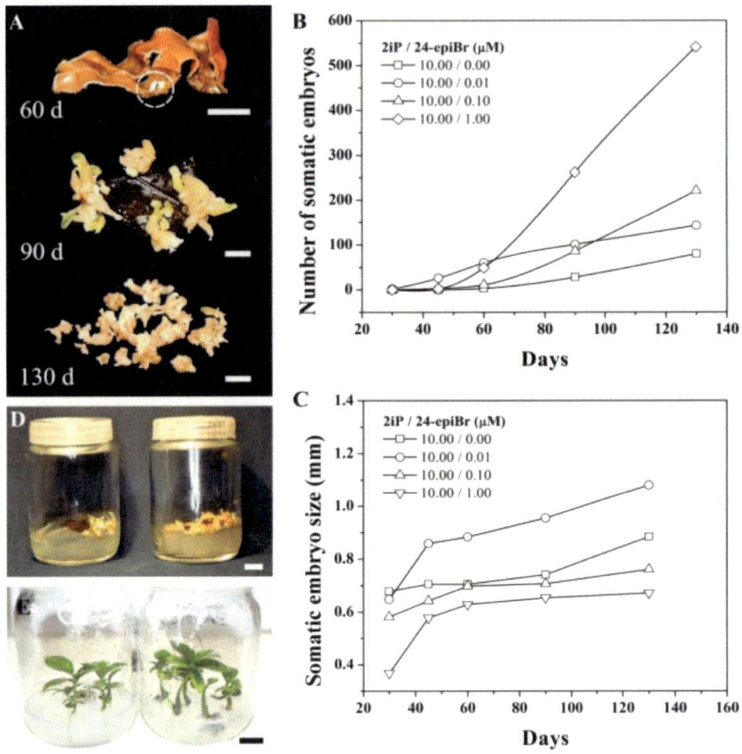

Figure 6.
Direct somatic embryogenesis from leaf explants of C. arabica. *(A) Macroscopic view of coffee leaf explants after 60, 90 and 130 days of culture on induction medium supplemented with 10 μM 2-iP + 1.0 μM 24-epiBR. (B) Number and (C) size of somatic embryos obtained by treatment during the somatic embryogenesis process. (D) Somatic embryo germination. (E) Elongating regenerated plantlets after 90 days on ½MS medium. Bars = 0.5 and 2 cm [99].*

has also been used to induce regeneration of *C. arabica* somatic embryos via the direct pathway [47, 48]. Leaf explants of cultivar IAPAR 59 and *C. arabica* hybrid Sachimor showed direct somatic embryogenesis response in the presence of TDZ at concentrations of 2.27; 4.54; 6.81; 9.08; 13.62 μM, but with low embryo production [96].

The literature shows that it is well established that the *Coffea* direct pathway only occurs in the presence of cytokine, but it is also possible to find studies in which explants of this species formed somatic embryos in the presence of auxin. Explants of cultivar Acaia Cerrado formed somatic embryos when grown in a single culture medium with addition of kinetin, GA_3 and NAA [67]. In another study, it was found that leaf explants of cultivar Mundo Novo submitted to direct pathway showed high production of somatic embryos in response to 2-iP treatment associated with brassinosteroid compared to 2-iP alone control [99]. On the other hand, explants treated with brassinosteroid alone without cytokine formed only embryogenic structures without any occurrence of somatic embryos (**Figure 6**).

4.3 Stress factor

The stress factors have been related to promoting the acquisition of embryogenic competence in different species [20, 66]. Stressful conditions can also influence the acquisition of embryogenic competence in different species [19, 20, 30, 66, 100, 101].

Figure 7.
Effect of 3 and 6 g/L agar on the direct somatic embryogenesis in explants of C. canephora *cultivar "Robusta 2264 Mar" maintained at 25°C in the dark [106]. (A) 6 g/L of agar: explants with curvature without contact with the medium. (B) Explant with few somatic embryos in the presence of 6 g/L agar. (C) 3 g/L of agar: the medium is in contact with the border of the explant. (D) Explant with many somatic embryos in the presence of 3 g/L agar.*

Studies indicate that the occurrence of somatic embryogenesis is strongly related to the exposure of explants to some high intensity stress factor and or the high concentration of plant growth regulator [22].

Osmotic stress treatments alter the explant's environment. This change in tissue/organ growth conditions may represent the estrum that enables cells to undergo changes in developmental processes and make them competent for inductive signals for somatic embryogenesis. Thus, the stress-induction system is composed of two phases: the acquisition of embryogenic competence and the formation of the somatic embryo [102].

Of the responses found in different species, it was shown that the stress-induction system could cause a greater formation of somatic embryos [101–107] although its forms of control and action are unknown and this aspect has been little studied in the culture of coffee plants. It was verified that *C. canephora* explants submitted to direct somatic embryogenesis formed a greater number of embryos in a medium to which 3 g of agar had been added, than in one containing 6 g [106] (**Figure 7**). This result was indirect evidence that altering the osmotic potential of the culture tends to favor the ability of somatic embryogenesis. In another study, it was found that the alteration of the osmotic concentration of the culture medium influenced the embryogenesis response [108]. The use of 7% PEG 6000 caused a greater formation of somatic embryos in foliar explants of the *C. arabica* genotypes AC1 and cultivar Mundo Novo than the use of 5% PEG 6000. This reagent has a high molecular weight and is inert, non-ionic, non-toxic, water soluble [109], not absorbed by vegetable cells and alters the osmotic potential when added to a culture medium.

5. Conclusions

Somatic embryogenesis contributes to coffee crop both in relation to breeding programs and its production chain. Little is known about the factors controlling somatic embryogenesis in *Coffea* genotypes. But it is known that plant hormones act in controlling the occurrence of this process. In addition, studies have shown that environmental and mainly stress factors applied during the cultivation condition are involved in the control of somatic embryogenesis in *Coffea*.

Nomenclature

MS	Murashige and Skoog
2,4 D	2,4-dichlorophenoxyacetic acid
2-iP	N6-(2-isopentenyl)adenine
NAA	naphthylacetic acid
6-BA	6-benzyladenine
Ki	kinetin
ZE	zeatin
TDZ	thidiazuron
PEG 6000	polyethyleneglycol 6000

Author details

Julieta Andrea Silva de Almeida
Instituto Agronômico de Campinas, Campinas, Brazil

*Address all correspondence to: julietasa@iac.sp.gov.br

References

[1] Davis AP, Govaerts R, Bridson DM, Stoffelen P. An annoted taxonomic conspectus of the genus *Coffea* (Rubiaceae). Botanical Journal of the Linnean Society. 2006;**152**:465-512. DOI: 10.1111/j.1095-8339.2006.00584.x

[2] Barbosa IP, Costa WG, Nascimento M, Cruz CD, Oliveira ACB. Recommendation of *Coffea arabica* genotypes by factor analysis. Euphytica. 2007;**215**:178. DOI: 10.1007/s10681-019-2499-x

[3] Carvalho A. Cruzamento entre *C. arabica* e *C. canephora*. In: Proceedings of the 10th Colloquium of International Coffee Science Association, ASIC, 11-14 October. Brasil: Salvador; 1982. pp. 363-368

[4] Etienne H, Breton D, Breitler J-C, Bertrand B, Déchamp E, Awada R, et al. Coffee somatic embryogenesis: how did research, experience gained and innovations promote the commercial propagation of elite clones from the two cultivated species? Frontiers in Plant Science. 2018;**9**:1-21. DOI: 10.3389/fpls.2018.01630

[5] Medina-Filho HP, Bordignon R, Carvalho CHS. Desenvolvimento de novas cultivares de café arábica. In: Carvalho CHS, editor. Cultivares de Café: Origem, Características e Recomendações. 1st ed. Embrapa Café: Brasília; 2008. pp. 79-101. ISBN: 978-85-61519-00-1

[6] Vieira LGE, Kobayashi AK. Micropropagação do cafeeiro. In: Anais do I Simpósio de Pesquisas dos Cafés do Brasil, 10-14 September. Minas Gerais: Poços de Caldas; 2000. pp. 147-167

[7] Corredoira E, Merkle SA, Martínez MT, Toribio MJM, Canhoto MJM, Correia SI, et al. Non-zygotic embryogenesis in hardwood species. Critical Reviews In Plant Sciences. 2019;**38**:29-97. DOI: 10.1080/07352689.2018.1551122

[8] Mendez-Hernandez HA, Ledezma-Rodriguez M, Avilez-Montalvo RN, Juarez-Gomez YL, Skeete A, Avilez-Montalvo J, et al. Signaling overview of plant somatic embryogenesis. Frontiers in Plant Science. 2019;**10**:1-15. DOI: 10.3389/fpls.2019.00077

[9] Covre AM, Partelli FL, Mauri AL, Dias MA. Crescimento e desenvolvimento inicial de genótipos de café conilon. Revista Agro@mbiente Online. 2013;**7**:193-202. DOI: 10.18227/1982-8470ragro.v7i2.944

[10] Bergo CL, Mendes ANG. Propagação vegetativa do cafeeiro (*Coffea arabica* L.) por meio de enraizamento de estacas. Ciência & Agrotecnologia. 2000;**24**:392-398. Available from: http://portalrev.enfermagem.bvs.br/index.php?issn=1413-7054&lang=pt

[11] Osuji C, Abubakar S, Mowobi G. The place of somatic embryogenesis in crop improvement through genetic transformation. Journal of Environment and Life Sciences. 2016;**1**:1-7. Available from: www.imedpharm.com/journals/index.php/jels

[12] Carman JG. Embryogenic cells in plant tissue cultures: Occurrence and behavior. In Vitro Cell Development Biology. 1990;**26**:746-753. DOI: 10.1007/BF02623615

[13] Dodemam VL, Kreis DG. Zigotic embryogenesis versus somatic embryogenesis. Journal Experimental Botany. 1997;**48**:1493-1509. DOI: 10.1093/jxb/48.8.1493

[14] Quiroz-Figueroa FR, Rojas-Herrera R, Galaz-Avalos RM, Loyola-Vargas VM. Embryo production

through somatic embryogenesis can be used to study cell differentiation in plants. Plant Cell, Tissue and Organ Culture. 2006;**86**:285-301. DOI: 10.1007/s11240-006-9139-6

[15] Krikorian AD, Berquam DL. Plant cell and tissue culture: The role of Haberlandt. Botanical Review. 1969;**35**:59-67. DOI: 10.1007/BF02859888

[16] Williams EG, Maheswaran G. Somatic embryogenesis: Factors influencing coordinated behaviour of cells as an embryogenic group. Annals of Botany. 1986;**57**:443-462. DOI: 10.1093/oxfordjournals.aob.a087127

[17] Merkle SA, Parrott WA, Flinn BS. Morphogenic aspects of somatic embryogenesis. In: Thorpe TA, editor. In Vitro Embryogenesis in Plants. Dordrecht: Kluwer Academic Publishers; 1995. pp. 155-203

[18] Zimmerman JL. Somatic embryogenesis: A model for early development in higher plants. The Plant Cell. 2010;**5**:1411-1423. DOI: 10.1105/tpc.5.10.1411

[19] Michaux-Ferrière N, Schwendiman J. Modalités d'initiation des cellules à l'origine des embryons somatiques. Acta Botanica Gallica. 1993;**140**:603-613. DOI: 10.1080/12538078.1993.10515641

[20] Fehér A, Pasternak TP, Dudits D. Transition of somatic plant cells to an embryogenic state. Plant Cell, Tissue and Organ Culture. 2003;**74**:201-228. DOI: 10.1023/A:1024033216561

[21] Jiménez VM. Involvement of plant hormones and plant growth regulators on *in vitro* somatic embryogenesis. Plant Growth Regulation. 2005;**47**:91-110. DOI: 10.1007/s10725-005-3478-x

[22] Fehér A. Somatic embryogenesis—Stress-induced remodeling of plant cell fate. Biochimica et Biophysica Acta (BBA)—Gene Regulator Mechanism. 2015;**1849**:385-405. DOI: 10.1016/j.bbagrm.2014.07.005

[23] Werner ET, Lima ABP, Amaral JAT. Expressão gênica na embriogênese somática vegetal. Enciclopédia Biosfera, Centro Científico Conhecer—Goiânia. 2012;**8**(14):552-580

[24] Siddiqui ZH, Abbas ZK, Ansari MW, Khan MN. The role of miRNA in somatic embryogenesis. Genomic. 2019;**111**:1026-1033. DOI: 10.1016/j.ygeno.2018.11.022

[25] Williams RR. The chemical mirroenviroment. In: Jenny-Aitken C, Kozai T, Smith L, editors. Automation and Environmental Control in Plant Tissue Culture. The Netherlands: Kluwer Academic Publisher; 1995. pp. 405-439

[26] Isah T. Induction of somatic embryogenesis in woody plants. Acta Physiology Plantarum. 2016;**38**:118. DOI: 10.1007/s11738-016-2134-6

[27] Conger BV, Hanning GE, Gray DJ, Mcdaniel JK. Direct embryogenesis from mesophyll cells of orchardgrass. Science. 1983;**221**:850-851. DOI: 10.1126/science.221.4613.850

[28] von Arnold S, Sabala I, Bozhkov P, Dyachok J, Filonova L. Developmental pathways of somatic embryogenesis. Plant Cell, Tissue and Organ Culture. 2002;**69**:233-249. DOI: 10.1023/A:1015673200621

[29] Haccius B. Question of unicellular origin of non-zygotic embryos in callus cultures. Phytomorphology. 1978;**28**:74-81. Available from: http://agris.fao.org/agris-search/search.do

[30] Gaj MD. Factors influencing somatic embryogenesis induction and plant regeneration with particular reference to *Arabidopsis thaliana* (L.) Heynh. Plant

Growth Regulation. 2004;**43**:27-47. DOI: 10.1023/B:GROW.0000038275.29262.fb

[31] Jiménez VM, Bangerth F. Endogenous hormone levels in explants and in embryogenic and non-embryogenic cultures of carrot. Physiologia Plantarum. 2001;**111**:389-395. DOI: 10.1034/j.1399-3054.2001.1110317.x

[32] Eira MTS, Silva EAA, Castro RD, Dussert S, Walters C, Derek BJ, et al. Coffee seed physiology. Brazilian Journal Plant Physiology. 2006;**18**:149-163. Available from: Http://www.Scielo.Br/Pdf/Bjpp/V18n1/A11v18n1.Pdf

[33] Souza NA, Rosa SDVF, Vilela AL, Figueiredo MA, Vilela ALO, Stphano-Filho R, et al. Big coffee Vl: Seed desiccation tolerance, sieve classification. Coffee Science. 2018;**13**:510-518. DOI: 10.25186/cs.v13i4.1492

[34] Söndahl MR, Sharp WR. High frequency induction of somatic embryos in cultured leaf explants of *Coffea arabica* L. Zeitschrift für Pflanzenphysiologie. 1977;**81**:395-408. DOI: 10.1016/S0044-328X(77)80175-X

[35] Staritsky G. Embryoid formation in callus cultures of coffee. Acta Botanica Neerlandica. 1970;**19**:509-514. DOI: 10.1111/j.1438-8677.1970.tb00677.x

[36] Altman A, Loberant B. Micropropagation: Clonal plant propagation *in vitro*. In: Altman A, editor. Agricultural Biotechnology. New York: Marker Dekker; 1998. pp. 19-42

[37] Kumar V, Naidu MM, Ravishankar GA. Developments in coffee biotechnology *in vitro* plant propagation and crop improvement. Plant Cell Tissue Organ Culture. 2006;**87**:49-65. DOI: 10.1007/s11240-006-9134-y

[38] Nic-Can GI, Galaz-Ávalos RM, De-la-Peña C, Alcazar-Magaña A, Wrobel K, Loyola-Vargas VM. Somatic embryogenesis: Identified factors that lead to embryogenic repression. A case of species of the same genus. PLOS One. 2015;**10**:1-21.e01264. DOI: 10.1371/jornal:pone.0126414

[39] Figueira ER, Londe LN, Marques RV, Marques SV, Silva AS, Luz JMQ. 2,4-D, silver nitrate, and acetylsalicylic acid influence on the *in vitro* culture of coffee (*Coffea arabica* L.) anthers. Científica. 2008;**36**(1):27-33. DOI: 10.15361/1984-5529.2008v36n1p27+-+33

[40] Silva AS, Luz JMQ, Rodrigues TM, Bittar CA, Lino LO. Callus induction and embryo regeneration in *Coffea arabica* L. anther by silver nitrate and ethylene. Revista de Ciência Agronômica. 2011;**42**:921-929. Available from: http://ccarevista.ufc.br/seer/index.php/ccarevista/article/view/850/627

[41] Sharp WR, Caldas LS, Crocomo OJ, Monaco LC, Carvalho A. Production of *Coffea arabica* callus of three ploidy levels and subsequent morphogenesis. Phyton. 1973;**31**:67-74. Available from: http://www.scielo.org.ar/cgi-bin/wxis.exe/iah/

[42] Dublin P. Techniques de reproduction végétative *in vitro* et amélioration génétique chez les cafeiers cultivés. Café Cacao Thé. 1984;**28**:231-244. Available from: http://agritrop.cirad.fr/11584/

[43] Baumann TW, Neuenschwander B. Tissue culture in coffee biotechnology. Café Cacao Thé. 1990;**34**:159-164. Available from: http://agritrop.cirad.fr/11584/

[44] Fernández-Da Silva R, Gallardo LH, Yuffá AM. Primary and secondary somatic embryogenesis in leaf sections and cell suspensions of *Coffea arabica*

cv. 'Catimor'. Interciencia. 2000;**30**:694-698. Available from: https://www.interciencia.net/

[45] Almeida JAS, Silvarolla MB, Fazuoli LC, Stancato GC. Embriogênese Somática em genótipos de *Coffea arabica* L. Coffee Science. 2008;**3**:143-151. DOI: 10.25186/cs.v3i2.85

[46] Almeida JAS, Silvarolla MB. Induction of somatic embryos of *Coffea arabica* genotypes by 6-benzyladenine. International Journal of Plant Developmental Biology. 2009;**53**:5-8. Available from: http://www.globalsciencebooks.info/Online/GSBOnline/OnlineIJPDB_3_1.html

[47] Kahia J, Kirika M, Lubabali H, Mantel S. High-frequency direct somatic embryogenesis and plantlet regeneration from leaves derived from in vitro-germinated seedlings of a *Coffea arabica* hybrid cultivar. Hortscience. 2016;**51**:1148-1152. DOI: 10.21273/Hortsci10771-16

[48] Yi-Chieh W, Meng-Ze L, Bin H, Hsiao-Hang C, Jen-Tsung C. Thidiazuron enhanced somatic embryogenesis from callus lines of arabica coffee and subsequent plant regeneration. Acta Biologica Cracoviensia Series Botanica. 2018;**60**:35-44. DOI: 10.24425/118053

[49] Bieysse D, Gofflot A, Michaux-Ferrière N. Effect of experimental conditions and genotypic variability on somatic embryogenesis in *Coffea arabica*. Canadian Journal Botany. 1993;**71**:1496-1502. DOI: 10.1139/b93-173

[50] Ribas AF, Pereira LFP, Vieira LGE. Genetic transformation of coffee. Brazilian Journal Plant Physiology. 2006;**18**:83-94. DOI: 10.1590/S1677-04202006000100007

[51] Gatica-Arias AM, Arrieta G, Espinoza AMV. Comparison of three *in vitro* protocols for direct somatic embryogenesis and plant regeneration of *Coffea arabica* L. cvs. caturra and catuaí. Agronomia Costarricense. 2007;**31**:85-94. DOI: 10.2225/vol11-issue1-fulltext-9

[52] Barry-Etienne D, Bertrand B, Schlönvoigt A, Etienne H. The morphological variability within a population of coffee somatic embryos produced in a bioreactor affects the regeneration and the development of plants in the nursery. Plant Cell Tissue Organ Culture. 2002;**68**:153-162. DOI: 10.1023/A:1013874221569

[53] Bonga JM, Klimaszewska KK, von Aderkas P. Recalcitrance in clonal propagation, in particular of conifers. Plant Cell Tissue Organ Culture. 2010;**100**:241. DOI: 10.1007/s11240-009-9647-2

[54] Almeida JAS, Silvarolla MB, Braghini MT, Fazuoli LC. Somatic embryogenesis in genotypes of *Coffea arabica* with reduced caffeine content. In: Proceedings of the 21th Colloquium of International Coffee Science Association, ASIC, 10-15 September. France: Montpellier; 2006. pp. 885-889

[55] Berthouly M, Michaux-Ferriére NM. High frequency somatic embryogenesis I *Coffea canephora*. Induction conditions and histological evolution. Plant Cell Tissue and Organ Culture. 1996;**44**:169-176. DOI: 10.1007/BF00048196

[56] Almeida JAS, Fazuoli LC, Ramos LCS. Efeito das estações do ano na calogênese e embriogênese somática em oito genótipos de *Coffea arabica*. In: Anais do IV Simpósio de Pesquisas dos Cafés do Brasil, Consórcio Brasileiro de Pesquisa e Desenvolvimento do Café, 2-5 May. Paraná: Londrina; 2005. pp. 1-5

[57] Santana-Buzzy N, Herrera RR, Ávalos RMG, Ku-Cauich JR, Mijangos-Cortés J,

Gutiérrez-Pacheco LC, et al. Advances in coffee tissue culture and its pratical applications. In Vitro Cell Development Biology-Plant. 2007;**3**:507-520. DOI: 10.1007/s11627-007-9074-1

[58] Larkin PJ, Scoweroft WR. Somaclonal variation—A novel source of variability from cell cultures for plant improvement. Theorical Applied Genetic. 1981;**60**:197-214. DOI: 10.1007/BF02342540

[59] Krishna H, Alizadeh M, Singh D, Singh U, Chauhan N, Eftekhari M, et al. Somaclonal variations and their applications in horticultural crops improvement. Biotech. 2016;**6**:54. DOI: 10.1007/s13205-016-0389-7

[60] Bolivar-Gonzalez A, Valdez-Melara M, Gatica-Arias A. Responses of Arabica coffee (*Coffea arabica* L. var. Catuai) cell suspensions to chemically induced mutagenesis and salinity stress under *in vitro* culture conditions. In Vitro Cellular & Developmental Biology-Plant. 2018;**54**:576-589. DOI: 10.1007/s11627-018-9918-x 2018

[61] Murashige T, Skoog F. A revised medium for rapid growth and bioassays with tobacco tissue cultures. Physiologia Plantarum. 1962;**15**:473-497. DOI: 10.1111/j.1399-3054.1962.tb08052.x

[62] Dublin P. Embryogenèse somatique directe sur fragments de feuilles de caféier arabusta. Café Cacao Thé. 1981;**25**:237-241. Available from: http://agritrop.cirad.fr/11584/

[63] Lanaud C. Production de plantules de *C. canephora* par embryogenèse somatique réalisée à partir de culture *in vitro* d´ovules. Café Cacao Thé. 1981;**25**:231-236. Available from: http://agritrop.cirad.fr/454057/

[64] Pierson ES, Van Lammeren AAM, Schel JHN, Staritsky G. *In vitro* development of embryoids from punched leaf discs of *Coffea canephora*. Protoplasma. 1983;**115**:208-216. DOI: 10.1007/BF01279811

[65] Sanglard NA, Silva PAM, Sattler MC, Oliveira SC, Cesário LM, Ferreira A, et al. Indirect somatic embryogenesis in *Coffea* with different ploidy levels: A revisiting and updating study. Plant Cell Tiss Organ Culture. 2019;**136**:255-267. DOI: 10.1007/s11240-018-1511-9

[66] Zavattiere MA, Frederico AM, Lima M, Sabino R, Arnholdt-Shmitt B. Induction of somatic embryogenesis as example of stress-relatec plant actions. Electronic Journal of Biotechnology. 2010;**13**:1-9. DOI: 10.2225/vol13-issue1-fulltext-4

[67] Pereira AR, Carvalho SP, Pasqual M, Santos FC. Embriogênese somática direta em explantes foliares de *Coffea arabica* L. cv. Acaiá cerrado: Efeito de cinetina e ácido giberélico. Ciência e Agrotecnologia. 2007;**31**:332-336. Available from: http://www.scielo.br/pdf/cagro/v31n2/a10v31n2.pdf

[68] Silva AS, Luz JMQ, Rodrigues TM, Marques SV, Marques RV, Pasqual M. *Coffea arabica* L. anther callus and pro-embryoid induction by different growth regulators. Bioscience Journal. 2009;**25**:19-27. Available from: http://www.seer.ufu.br/index.php/biosciencejournal/article/view/6946

[69] Cid LPB, Cruz ARR. Somatic embryogenesis from three coffee cultivars: 'Rubi', Catuaí Vermelho 81 and IAPAR 59. HortScience. 2004;**39**:130-131. DOI: 10.21273/HORTSCI.39.1.130

[70] Santana N, Gonzalez ME, Valcarcel M, Canto-Flick A, Hernandez M, Fuentes-Cerda CFJ, et al. Somatic embryogenesis: A valuable alternative for propagating selected robusta coffee (*Coffea canephora*) clones. In vitro Cellular and Developmental Biology—Plant. 2004;**40**:95-101. DOI: 10.1079/IVP2003486

[71] Bartos PMC, Gomes HT, Amaral LIV, Teixeira JB, Scherwinski-Pereira JE. Biochemical events during somatic embryogenesis in *Coffea arabica* L. Biotech. 2018;**8**:209a. DOI: 10.1007/s13205-018-1238-7

[72] Bartos PMC, Gomes HT, Gomes SM, Vasconcelos Filho SC, Teixeira JB, Scherwinski-Pereira JE. Histology of somatic embryogenesis in *Coffea arabica* L. Biologia. 2018;**1255b**:73. DOI: 10.2478/s11756-018-0131-5

[73] Yeung EC. Structural and developmental patterns in somatic action. In: Thorpe TA, editor. In Vitro Embryogenesis in Plants. The Netherlands: Kluwer Academic Publishers; 1995. p. 205-248. ISBN: 978-94-011-0485-2

[74] Bandel G, Carvalho FJPC, Crocomo OJ, Sharp WR, Gutierrez LE, Carvalho PCT. Aspectos citológicos da diferenciação de tecidos de cafeeiro cultivados *in vitro*. In: Anais da Escola Superior de Agronomia "Luiz de Queiroz". Vol. 32. 1975. pp. 717-723. DOI: 10.1590 / S0071-12761975000100064

[75] Almeida JAS, Leal RR, Carmazini VCB, Salomon MV, Guerreiro-Filho O. Characterization of the morphological events in the somatic embryogenesis in the direct somatic of *Coffea arabica* L. genotypes. Plant Cell Biotechnology and Molecular Biology. 2016;**17**:393-403. Available from: https://journals.indexcopernicus.com/search/article?articleId=2035032

[76] McDaniel CN. Competence, determination and induction in plant development. In: Malacinski GM, Bryant SV, editors. Pattern Formation a Primer in Developmental Biology. New York: Macmillan Publishing; 1984. pp. 393-412

[77] Sharp WR, Sodhal MR, Evans AE, Caldas LA, Maraffa SB. The physiology of in vitro asexual embryogenesis. Horticultural Reviews. 1980;**2**:268-310

[78] Wann SR. Somatic embryogenesis in woody species. Horticultural Reviews. 1988;**10**:153-181. DOI: 10.1002/9781118060834.ch5

[79] Maciel ALR, Pasqual M, Pereira AB, Silva AB, Dutra LF. Embriogênese somática indireta em explantes foliares de *Coffea arabica* cv. Obatã. Ciência e Agrotecnologia. 2003;**27**:107-116. DOI: 10.1590/S1413-70542003000100013

[80] Chée RP, Cantliffe DJ. Selective enhancement of *Ipomea batatas* Poir. Embryogenic and non-embryogenic callus growth and production of embryos in liquid culture. Plant Cell, Tissue and Organ Culture. 1988;**15**: 149-159. DOI: 10.1007/BF00035756

[81] Raghavan V. Molecular Embryology of Flowering Plants. 1st ed. New York: Cambridge University Press; 1997. 690 p

[82] Emons AMC. Somatic embryogenesis: cell biological aspects. Acta Botanica Neerlandica. 1994;**43**:1-14. DOI: 10.1111/j.1438-8677.1994.tb00729.x

[83] Motoike SY, Saraiva ES, Ventrella MC, Silva CV, Salomão LCC. Somatic embryogenesis of *Myrciaria aureana* (Brazilian grape tree). Plant Cell Tissue Organ Culture. 2007;**89**:75-81. DOI: 10.1007/s11240-007-9210-y

[84] Michaux-Ferriére N, Bieysse D, Alvard D, Dublin P. Étude histologique de l`embryogenêse somatique chez *Coffea arabica*, induite par culture sur milieux uniques de fragments foliaires de génotypes différents. Café Cacao Thé. 1989;**33**:207-217. DOI: 10.1007/BF00048196

[85] West MAL, Harada JJ. Embryogenesis in higher plants: an overview. Plant Cell. 1993;**5**:1361-1369. DOI: 10.1105/tpc.5.10.1361

[86] Street HE. Embryogenesis and chemically induced organogenesis. In: Sharp WP, Larsen PO, Paddock EF, Raghavan V, editors. Plant Cell and Tissue Culture Principles and Applications. 4th ed. Columbus: Ohio State University Press; 1977. pp. 123-153. ISBN: 081420287X/9780814202876

[87] Ramos LCS, Yokoo EY, Gonçalves W. Direct somatic embryogenesis is genotype specific in coffee. In: Proceedings of the 15th Colloquium of International Coffee Science Association, ASIC, 24-30 July. France: Montpellier; 1993. pp. 763-766

[88] Halpering W. Population density effects on embryogenesis in carrot cell cultures. Exp Cell Research. 1967;**48**:170-173. DOI: 10.1016/0014-4827(67)90292-3

[89] Boutchouang RP, Akitio OFZ, Tchouatcheu AGN, Niemenak N. Influence of the position of flowers buds on the tree on somatic embryogenesis of cocoa (*Theobroma cacao* L.). International Journal of Plant Physiology and Biochemistry. 2016;**8**:7-16. DOI: 10.5897/IJPPB2016.0247

[90] Almeida JAS, Leal RR, Carmazini VCB, Salomon MV, Guerreiro-Filho O. Effect of temperature and cytokinin on the capacaity of somatic embryogenesis in *Coffea arabica* L. genotypes. Coffee Science. 2014;**9**:394-399. DOI: 10.25186/cs.v9i3.668

[91] Yasuda T, Fujii Y, Yamaguchi T. Embryogenic callus induction from *Coffea arabica* leaf explants by benzyladenine. Plant Cell Physiology. 1985;**26**:595-597. DOI: 10.1093/oxfordjournals.pcp.a076946

[92] Hatanaka T, Azuma T, Uchida N, Yasuda T. Effect of plant hormones on somatic embryogenesis of *Coffea canephora*. In: Proceedings of the 16th Colloquium of International Coffee Science Association, ASIC, 9-14 April. Japan: Kyoto; 1995. pp. 790-797

[93] Acuña MEA. Somatic embryogenesis induced by culture on single media in coffee plants from crosses of *Coffea arabica* by Timor hybrid. In: Proceedings of the 15th Colloquium of International Coffee Science Association, ASIC, 24-30 July; Montpellier. 1993. pp. 790-889

[94] Vasconcellos RCC, Almeida JAS, Silvarolla MB. Indução de embriões somáticos em explantes foliares de genótipos de *Coffea arabica* em presença da citocinina 2-iP. In: Anais do VI Simpósio de Pesquisas dos Cafés do Brasil, Consórcio Brasileiro de Pesquisa e Desenvolvimento do Café, 2-5 June. Vitória: Espírito Santo; 2009. ISSN: 1984-9245. pp. 1-5

[95] Alves IS, Carmazini VCB, Santos CD, Almeida JAS. 2-Isopentenyladenine in the induction of direct somatic embryogenesis capacity of *Coffea arabica* L. Ciência Rural. 2018;**48**:1-5. DOI: 10.1590/0103-8478cr20180001

[96] Ayub RA, Gebieluca AN. Embriogênese somática em genótipos de café (*Coffea arabica*) é citocinina dependente. Publicação Universidade Estadual de Ponta Grossa Ciências Exatas, da Terra, Ciêcias Agronômica, Engenharia. 2003;**9**:25-30. DOI: 10.5212/publicatio.v9i02.806

[97] Rojas-Lorz L, Arrieta-Espinoza G, Valdez-Melara M, Pereira LFP, Gatica-Arias A. Influence of silver nitrate on somatic embryogenesis induction in Arabica Coffee (*Coffea arabica* L.). Brazilian Archives

of Biology and Technology. 2019;**62**:e19180228. DOI: 10.1590/1678-4324-2019180228

[98] Almeida JAS. Cultura de tecido em *Coffea*. In: Thomaziello RA, Fazuoli LC, editors. IX Curso de Atualização em Café, Documento IAC 91. 1st ed. Instituto Agronômico: Campinas; 2009. pp. 63-67

[99] Chone RMS, Rocha DIR, Monte-Bello CC, Pinheiro HP, Dornelas MC, Haddad CRB, et al. Brassinosteroid increases the cytokinin efficiency to induce direct somatic embryogenesis in leaf explants of *Coffea arabica* L. (Rubiaceae). Plant Cell Tissue and Organ Culture. 2018;**35**:63-71. DOI: 10.1007/s11240-018-1443-4

[100] Bressan RA, Hasegawa PM, Handa AK. Resistance of cultured higher plant cells to polyethylene glycol-induced water stress. Plant Science Letters. 1981;**21**:23-30. DOI: 10.1016/0304-4211(81)90065-1

[101] Bressan RA, Handa AK, Handa S, Hasegawa PM. Growth and water relations of cultured tomato cells after adjustment to low external water potentials. Plant Physiology. 1982;**70**:1303-1309. DOI: 10.1104/pp.70.5.1303

[102] Kikuchi A, Sanuki N, Higashi K, Koshiba T, Kamada H. Abscisic acid and stress treatment are essential for the acquisition of embryogenic competence by carrot somatic cells. Planta. 2006;**223**:637-645. DOI: 10.1007/s00425-005-0114-y

[103] Mohamed FM, Tawfik AA. Dehydration-induced alterations in growth and osmotic potential of callus from six terapy bean lines varying in drought resistance. Plant Cell Tissue Organ Culture. 2006;**87**:255-262. DOI: 10.1007/s11240-006-9159-2

[104] Ferrie AMR, Keller WA. Optimization of methods for using polyethylene glycol as a non-permeating osmoticum for the induction of microspore embryogenesis in the *Brassicacea*. In Vitro Cellular & Developmental Biology-Plant. 2007;**43**:348-355. DOI: 10.1007/s11627-007-9053-6

[105] Desjardins Y, Dubuc JF, Badr A. In Vitro Culture of Plants: a stressful activity. Acta Horticulturae (Ishs). 2009;**812**:29-50. Available from: www.Actahort.Org/Books/812/812

[106] Almeida JAS, Carmazini VCB, Ramos LCS. Indirect effect of agar concentration on the embryogenesis responses of *Coffea canephora*. Fruit, Vegetable and Cereal Science and Biotecnology. 2007;**1**:121-125. Available from: http://www.globalsciencebooks.info/Online/GSBOnline/OnlineFVCSB_1_2.html

[107] Morini S, D'Onofrio C, Fisichella M, Loreti F. Effect of high and low temperature on the leaf regenerating capacity of Quince BA29 rootstock. Acta Horticulturae. 2004;**658**:591-597. DOI: 10.17660/ActaHortic.2004.658.89

[108] Almeida JAS, Machado DFSP, Silvarolla MB, Machado EC. Effect of PEG on the somatic embryogenesis of *Coffea arabica* genotypes. In: Proceedings of the 22nd Colloquium of International Coffee Science Association, ASIC, 14-19 September. Brazil: Campinas; 2008. pp. 1020-1023

[109] Ahmad MAS, Javed F, Ashraf M. Iso-osmotic effect of NaCl and PEG on growth, cations and free proline accumulation in callus tissue of two indica rice (*Oryza sativa* L.) genotypes. Plant Growth Regulation. 2007;**53**:53-63. DOI: 10.1007/s10725-007-9204-0

Section 2

Coffee Growing

Chapter 2

Soil Management and Water-Use Efficiency in Brazilian Coffee Crops

Bruno Montoani Silva, Geraldo César de Oliveira, Milson Evaldo Serafim, Carla Eloize Carducci, Érika Andressa da Silva, Samara Martins Barbosa, Laura Beatriz Batista de Melo, Walbert Junior Reis dos Santos, Thiago Henrique Pereira Reis, César Henrique Caputo de Oliveira and Paulo Tácito Gontijo Guimarães

Abstract

Brazil is a world leader in coffee production. However, currently, it coexists with recurrent and severe droughts, accompanied by intense heat, strong insolation and low relative humidity. As the cultivation is carried out primarily in the rainy season, these world climate variations have affected crops yields and fruits quality, requiring innovative actions that promote efficient use of water stored in the soil. Among several soil management practices that promote a more rational use of water, deep tillage combined with liming, gypsum and fertilizer amendments lead to an increase in effective depth of coffee roots, therefore reducing water stress. Moreover, intercropping with *Urochloa* sp. is highly efficient in enhancing soil structure, water infiltration and plant available water capacity. Additionally, other innovative techniques and practices are also introduced in this chapter.

Keywords: soil water, soil structure, deep rooting, soil amendments, deep tillage, intercropping

1. Introduction

Adequate soil physical and hydrological conditions are key conditions for full plant development, which is a premise for coffee quality exportation due to requirements for grain quality and crop uniformity [1]. However, in the main coffee producing region of Brazil, there have been severe droughts. Although soils are mostly deep and able to store a large volume of water, they present small effective depth for the development of the root system, resulting in the edaphic drought, which has brought many losses to coffee farmers. This situation is aggravated in soils of oxide mineralogy and with very small granular structure, which condition the formation of pores with extreme diameters [2]. Thus, it leads to very rapid loss of water stored in very large pores, or to its strong retention in extremely small pores.

A number of measures have been sought by Brazilian researchers to solve the problems, such as selecting drought tolerant plants [3, 4]. However, a measure that has attracted the attention of most producers is the adoption of soil management systems that provide the best development of the root system of coffee crops and physical-hydrological adequacy of the soil.

Therefore, this chapter will discuss the main limitations of soils used in the main coffee growing area of Brazil and the mitigation techniques for soil suitability based on research that have been developed for over a decade.

2. Soil adequacy

2.1 Preparation, planting corrections and rooting

In the past, coffee cultivation was traditionally performed in grooves of 0.40 m × 0.40 m × 0.40 m. From the 1970s, the use of furrows for planting coffee was introduced in large scale. These furrows, open with tractors and with small furrowers, were shallow at 0.30–0.40 m deep and V-shaped, with small width at the bottom. For these reasons, and also due to the fact that under conventional coffee growing conditions soil fertilization was performed on the surface layers, much of the root system was limited to the first 0.40 m depth [5].

With the advancement of knowledge and technologies, it has been found that coffee roots can reach depths well above 1 m when in the absence of physical limitations and when adequate chemical conditions [6–8], such as sufficient calcium, phosphorus and boron contents, are provided [9].

With the development of new soil preparation tools, coffee farmers have been adopting deep furrow associated with soil correction and/or fertilization [10]. In the south and southwest regions of the state of Minas Gerais, deep tillage has often been carried out, allowing the incorporation of phosphate or limestone to a depth of 0.90 m. Due to higher soil turnover, larger amounts of fertilizer can be added in the furrow, correcting the soil in deeper layers and providing a better environment for coffee root development [6, 7, 10, 11].

Coffee cultivation using deep tillage system associated with surface application of additional doses of gypsum presents better drought resistance when compared with crops planted using conventional system, which conditions the permanence of the root system on the soil surface. Regarding the additional operation costs, the practice of deep tillage is compensated by the high crop yields in the first harvest [12]. Nevertheless, there are large variations in production costs, especially considering the price of the product.

Gypsum ($CaSO_4.2H_2O$) is considered a good soil conditioner due to its high mobility in the soil profile, providing calcium and sulfur to the plants, as well as acting as a deep corrective for toxic aluminum (**Figure 1**) [13].

The ability of gypsum to increase Ca^{2+} levels in the deepest soil layers is important for the proper development of the crop root system, especially because Ca^{2+} is the main component of the cell wall, being responsible for root elongation and growth [6–8].

The increase in effective CEC of the subsurface layers in management systems in which gypsum is applied is due to the increase in soil organic matter (SOM) (**Figure 2**). Coffee is mostly grown intercropped with Brachiaria between rows [10]. This grass is periodically mowed and its residues remain in the coffee line, representing continuous input of organic matter to the soil [8]. Thus, SOM contributes to

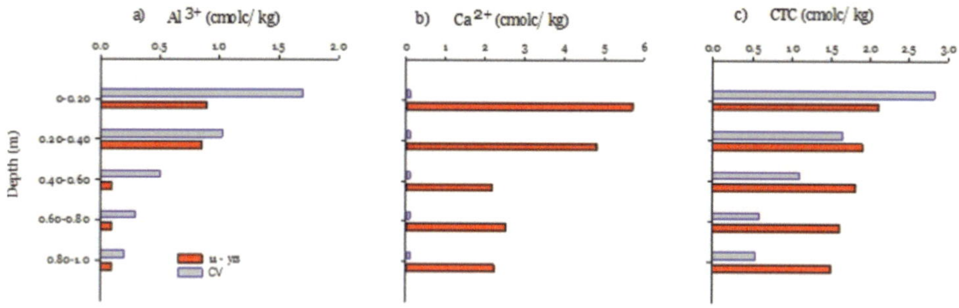

Figure 1.
Contents of Al^{3+} (a), Ca^{2+} (b) and effective CEC (c) in red Latosol under conventional management and after 11 years of deep preparation conservation system and application of 28 Mg ha^{-1} of gypsum. Source: From authors.

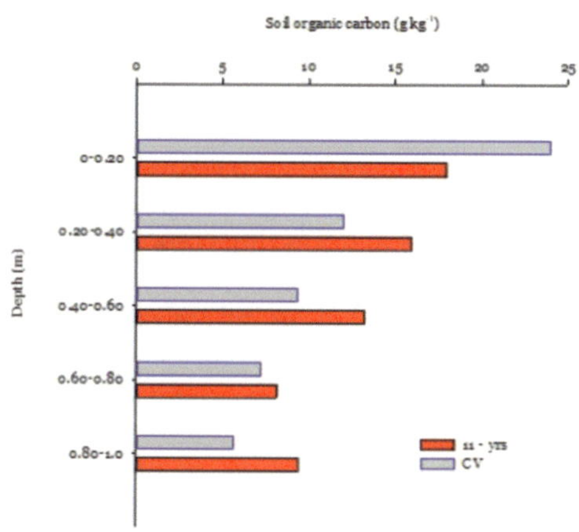

Figure 2.
Average values of organic carbon contents (g kg^{-1}) in red Latosol under conventional management and after 11 years in deep preparation conservation system and application of 28 Mg ha^{-1} of gypsum. Source: From authors.

increased CEC and improved nutrient utilization efficiency by providing a significant number of binding sites for essential elements present in the soil [14].

Studies showed that up to 21% of the carbon added by the roots could be incorporated into SOM [15]. Thus, the biomass of the coffee root system itself, favored by calcium, is also a source of organic matter for the soil and certainly contributed to raise the CEC (**Figure 1**).

Moreover, this management system can be considered efficient in the construction of fertility of Latosols, whose mineralogy is dominated by low chemical activity clay minerals (kaolinite and iron and aluminum oxides in the form of goethite, hematite and gibbsite). In these soils, which are typical of the Brazilian Cerrado biome, organic matter can contribute to up to 80% of negative soil loads [16].

Due to intense soil revolving, tillage management systems promote aggregate breakage, leading to significant structural changes [10, 12, 17]. However, by evaluating a Cambisol after 6 months of implantation of the coffee crop, Serafim et al. [17] observed a reduction in soil density and an increase in total porosity due to the benefits conditioned by the structural relief and construction of soil fertility. Serafim et al. [10] described the presence of coffee root system with average depth in the soil profile of 0.80 and 0.60 m at 6 months after planting for Latosol and Cambisol, respectively. After 1 year, the root system reached 1.40 m in Latossol and 1.20 m in Cambisol.

Serafim et al. [18], using the Least Limiting Water Range (LLWR) technique, found that a Cambisol presented no physical-water limitations after 3.5 years of coffee plantation and the crop implanted in this soil reached productivity much higher than the average of the state of Minas Gerais. It evidences the longevity of the positive effects of deep tillage on soil physical properties. Moreover, Serafim et al. [10, 17–19] and Silva [20] observed positive responses in soil physical properties, such as increase in the volume of large macropores (>147 μm), fine macropores (147–73 μm) and large mesopores (73–49 and 49–29 μm), when evaluating the physical quality of this soil after 5 years of tillage implementation. Similarly, Silva et al. [21] observed a significant increase in LLWR and a significant reduction in soil density when evaluating the structural quality of very clayey Latosol after 2 years of coffee cultivation.

Silva et al. [8] found a significant volume of inter-aggregate pores (macropores) after 3 years of coffee cultivation in Latosols, confirming the benefits of the management system using deep preparation associated with surface gypsum application. In the layer between 0.20–0.40 m of the soil, even after 5 years of cultivation, Silva [20] also found that soil management favored the expressive increase of pore volume of classes 9.0–2.9, 2.9–0.6 and 0.6–0.2 μm (mesopores), which is relevant since a good portion of the water retained in the soil will be available to the plants.

Particularly in Latosols under this management system, it was observed that in the absence of chemical and physical limitations of the soil the coffee root system reached depths greater than 1 m at 3 years of age (**Figure 3**), which is of fundamental importance to ensure crop survival in periods of edaphic drought [8].

Serafim et al. [19] evidenced intense water deficit up to 1.60 m in the crop line, when monitoring moisture of a very clayey and oxidic Latosol with 3.5 years of cultivation in a dry year. The authors attributed the results to the presence of roots that used intensely the available water in this depth of soil. Very thin roots were found in the soil layer between 1.50 and 1.70 m, indicating potential for water use in these deeper soil layers.

Similarly, in Cambisol, Serafim et al. [19] also showed more intense drying in the crop line up to 1.6 m caused by the roots of the plants, since active roots were found in this depth. The authors reported that although the crop does not have water availability in the layers closer to the surface in the dry period of the year, the larger volume of soil explored by the roots contributed to reduce the water deficit.

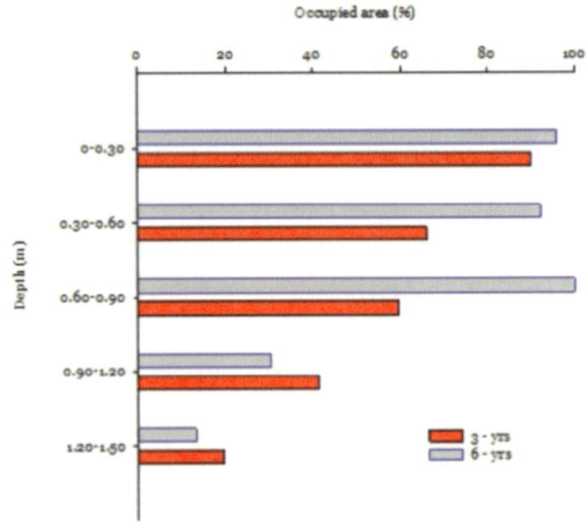

Figure 3.
Area occupied by coffee plant roots along the profile of Rhodic Haplustox. Source: Adapted from Silva et al. [8].

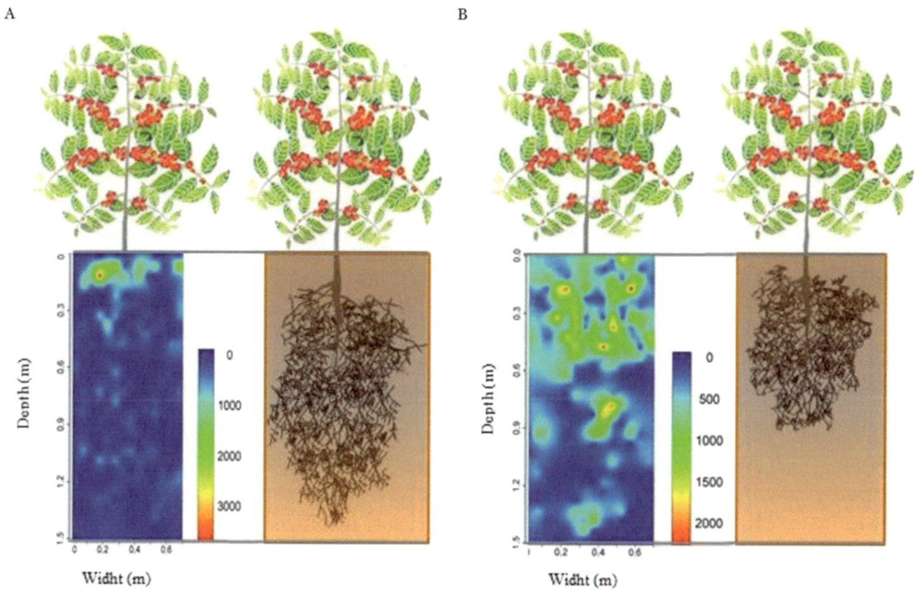

Figure 4.
Root length distribution (mm) in (A) gibbsitic Acrustox and (B) kaolinitic Haplustox both with the multipractice conservation management system for the coffee crop. Source: Adapted from Carducci et al. [7].

Given the above, it is noteworthy that although the benefits of mechanical soil revolving are readily apparent in coffee cultivation after 4 years of management [6, 7, 10, 17–19], studies show that these effects do not last long in some soil classes [22]. In this sense, particularly when soil is revolved, physical improvements to the soil may be temporary, since the durability of the changes depends on the texture and mineralogy of the soil [23].

Silva [20] reported that the deep tillage and gypsum management system was not effective in providing improvements in the physical quality of a Nitisol, since in the 0.0–0.20 m and 0.40–0.60 m layers, management provided a decrease in the volume of large macropores (>145 μm), which may affect the internal drainage of the profile. According to the author, in soils presenting textural B-horizon, the physical conditioning provided by soil preparation is short and the soil tends to reconsolidate. It is possible that clay illuviation may be acting in this process, as observed in Argisol by Marcolan and Anghinoni [24]. When soils are prepared there is a breakdown of aggregates, and an increase in soil clay dispersion [25].

Still regarding the development of the root system in Latosols, the practices employed in the management system described by Serafim et al. [10] also contributed to the coffee root growth, even in young (<3 years) roots [7], which are responsible for rapid water absorption and increased nutrient acquisition [26] (**Figure 4**).

The better distribution of the coffee root system in Latosol with high levels of gibbsite was promoted not only by the employed management system but also by the good distribution of well-connected pore diameters typical of this soil class (**Figures 4** and **9**). In kaolinitic Latosol, the system promoted the relief of the denser original structure, formed by thin and elongated pores promoted by the kaolinite mineral [27, 28], due to deep revolving associated with the addition of organic matter and gypsum, which favored concentrated root growth up to 0.80 m, but with regular root expansion with 500 mm length to 1 m depth (**Figure 4**) [7].

A well-distributed coffee root system along the soil profile, as observed in **Figure 4**, enhances the use of stored water available at greater depths (>0.80 m). Serafim et al. [19] and Silva et al. [29] reported the possibility of more efficient water absorption, minimizing the effects of water stresses to which these plants are subjected when cultivated in soils from the Cerrado biome, without harming crop yields [21]. Thus, knowledge about the distribution of coffee root system, as well as the probable changes in soil structure is the result of the interaction between the management system and the edaphoclimatic conditions that are intrinsic to Latosols.

2.2 Coffee intercropped with Brachiaria

The proper management of soil corrections and conditioning, dose adjustments and phosphorus use by the system, as well as balance in nutrient supply and leaf analysis for monitoring coffee nutrition are the main challenges of modern and competitive coffee cultivation for better use of available water in the soil–plant system [30]. Therefore, it is necessary to build soil fertility for sustainable coffee production in order to obtain increased nutrient use efficiency, increased fertilizer recovery rate, reduced biennial bearing and higher yield.

Coffee cultivation intercropped with *Brachiaria* (*Urochloa* sp.) improves the soil profile fertility. With vegetative intensification, the root system of the main crop naturally tends to deepen, accessing more water and nutrients, incorporating more carbon into the soil and improving its physical and biological quality [31]. In general, Brachiaria species have been considered prominent options for the production of plant residues to be incorporated in the soil or in its surface in no-tillage system, due to the good dry mass production and the high C/N ratio [32, 33]. In the intercropping system with coffee in low fertility soils, this behavior should also contribute to the increase of the soil organic matter (SOM) and consequently its cation exchange capacity (CEC), indirectly increasing the soil nutrients. *Urochloa ruziziensis* stands out among the species of Brachiaria, and is preferred by coffee growers because of its single flowering and well-developed root system with excellent field results [34].

The part of the coffee root system responsible for the absorption of water and nutrients, the thinnest roots, usually deepens to a depth of 40 cm [5] (**Figures 5** and **6**).

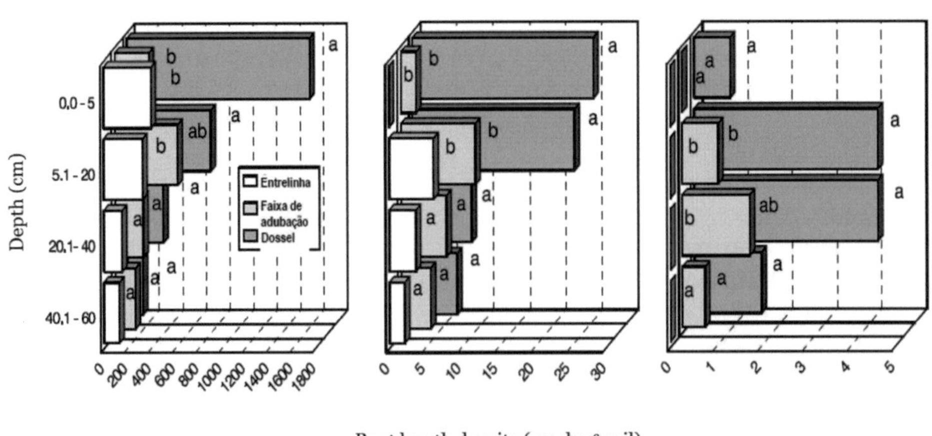

Figure 5.
Density of coffee roots as a function of the sampling site, below the canopy, below the fertilizer range, and in the center of the row. Source: Adapted from Motta et al. [5].

After a few years of planting under sufficient fertilizer application, some soil-moving nutrients such as nitrogen (N), potassium (K), sulfur (S) and boron (B) can leach beyond these absorbing roots. Thus, intercropping with deep-rooted plants practically all year round returns these nutrients to the surface of the soil–plant system. Therefore, managing between rows that collaborate with the proper management of soil fertility will certainly provide higher yields of coffee crop [35] due to the higher nutritional efficiency of the system production.

In addition, Brachiaria presents a root system that complements the efficiency of soil fertility use in the intercropping with the coffee as they explore depths of up to nearly 5 m (**Figure 7**).

In coffee cultivation intercropped with Brachiaria, plant residues are recycled and used as nutrients for coffee nutrition. The amount and regularity of plant residue addition is more important than the synchronization between release and nutrient demand by coffee because the increase in organic matter content over the years.

Brachiaria is more efficient than the coffee tree to extract the phosphorus from the soil, which will be available gradually with the decomposition/mineralization of the straw in the canopy projection. Over the years, the grass also incorporates this nutrient in depth as its root system develops in a larger volume of soil (**Figure 7**).

It is possible to estimate three plant cuts per cycle, with 5 tons of dry matter per hectare in each field based on Brachiaria average productivity data [37] and

Figure 6.
Root system (A) and aerial part (B) of productive coffee, with good management of soil fertility construction in association with Brachiaria. Photo: Paulo T. G. Guimarães.

Figure 7.
Root system of Brachiaria (Urochloa ruziziensis) pasture. (A) Detail of trench opening; (B) frontal view of Brachiaria roots; (C) view of Brachiaria roots from within the trench; (D) measurement of Brachiaria root system depth up to 4.9 m soil depth. Source: Revista Cafeicultura [36].

proportional adjustment of its soil exploration area in consortium with the coffee tree (up to 30% of the area). The nutritional contents in dry matter for each coffee brush operation are: 75 kg of N; 20.6 kg P_2O_5; 193 kg of K_2O; 24.4 kg of CaO; 20.8 kg of MgO; 3.5 kg of $S\text{-}SO_4$; 90 g of B; 55 g of Cu; 1 kg of Fe; 475 g of Mn, and 400 g of Zn [34]. For the availability of these nutrients in the crop cycle, it is necessary to mineralize the dry matter, which depends on the presence of water, temperature and microorganisms in the soil, since some nutrients, such as N and P, are partially released over a period of 3 years [38].

Despite the many advantages presented by the cultivation of Brachiaria between coffee lines, there may be some disadvantages, especially if the coffee grower

Figure 8.
Appropriate management of rows of coffee plants with Brachiaria after mowing. Photo: Geraldo C. Oliveira.

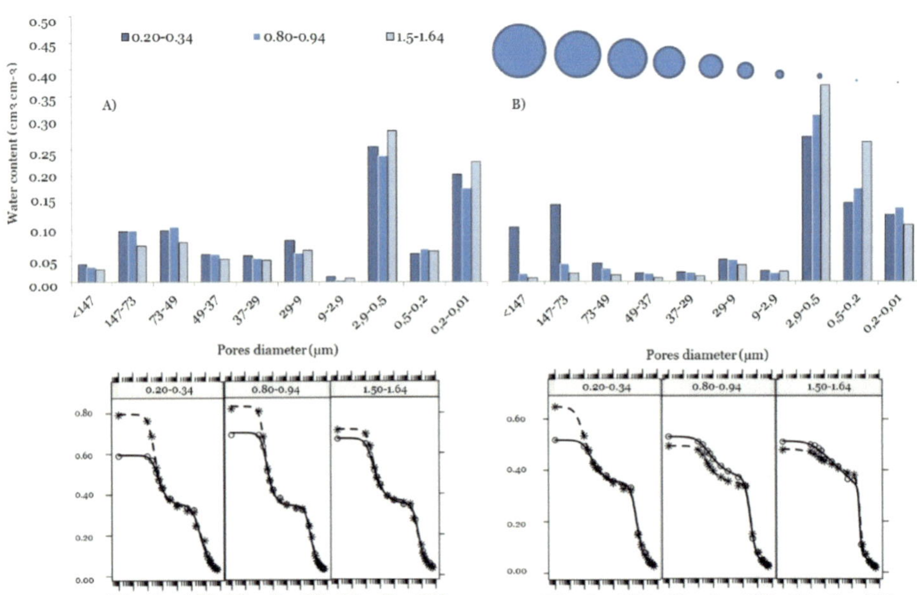

Figure 9.
Pore distribution of (A) gibbsitic Acrustox and (B) kaolinitic Haplustox, both with the multi-practice conservation management system for coffee cultivation at 0.20–0.34, 0.80–0.94 and 1.50–1.64 m depth. The pore diameter was extract of soil water retention curve by double van Genuchten model. Pore size <0.01 µm corresponds to >3500 kPa by WP4-T psychrometer [2]. The blue spheres represent the pore diameters. Source: From authors.

handles it incorrectly. Under conditions of severe water deficit, there may be competition for water and nutrients, harming the crop of commercial interest [39]. There may also be competition for nutrients and light and it is recommended to provide adequate and balanced coffee nutrition, as well as to maintain a strip of about 0.40 m on each side of the coffee trees, free from competing plants, and covered by residues from Brachiaria (**Figure 8**).

Coffee cultivation intercropped with Brachiaria is one of the practices of building soil fertility in profile for greater sustainability of coffee growing. The addition of this grass to the cultivation system is necessary for greater use of water and soil nutrients, which also allows the suppression of other difficult to control weeds, presenting several benefits for better coffee development and productivity and consequently greater profitability.

3. Porosity, water retention and availability in soils cultivated with coffee

The presence of an ideal pore network with a wide range of diameters is one of the key factors for high crop yields, especially those most demanding for water, such as coffee [21, 40]. Soil pore diameter and distribution interfere with drainage ratios, available water content, ion adsorption, root growth, aeration and temperature, acting directly on physical-water phenomena, being an indicator of soil quality [41–43].

Since soil mineral composition influences pore shape, length and connectivity, soils of oxide mineralogy, such as the very weathered Cerrado Latosols, tend to have a very strong, well-connected microgranular structure with large pore formation. There is formation of thinner and elongated pores [2, 27, 28, 43–45], which has implications on the water content available to plants.

When used in some production process such as food, fiber or energy, some structural change must occur, modifying the distribution and connection of their pore networks and, consequently, promoting changes in the soil air-water dynamics. In this sense, conservation agriculture [13] has as its principle the physical and chemical improvements of the root environment, by reducing soil tillage and maintaining living or dead surface cover. Thus, it minimizes the compressive and erosive processes, in addition to the oxidation of organic material, promoting the vertical growth of the root system of crops.

With these simple conservationist measures, coupled with the chemical corrections of acidic Latosols, improvements in the physical environment are expected, favored by the good development of the coffee roots, particularly by the reduction of restrictive impediments to the vertical growth of its roots and access to stored water [6–8, 29] (**Figure 9**).

Thus, the conservationist soil management system described by Serafim et al. [10] promoted changes in water retention in very weathered Latosols. According to Carducci et al. [2], the system was able to alter pore scaling such that it increased in the layer of 0–0.20-0.34 m the volume of large macropores (>147 μm) in kaolinitic Latosol and increased the intermediate diameters (73–2.9 μm), which are pores responsible for the gradual release of water to plants [43, 46]. There was also no limitation to aeration in soils (>147 μm: ≈ 15%), because the values were within the acceptable range for gas exchange maintenance (**Figure 9**).

According to Carducci et al. [2, 47, 48], genetically weathered Latosols present a large amount of interconnected structural pores, which facilitate drainage. Textural pores (including cryptopores) are responsible for water retention of high energy

[2, 43, 46, 49] However, because it was submitted to the conservationist management system, there was a small increase in the intermediate pores when compared to the greater depth evaluated in both soils, especially the one with gibbsite.

There is higher water retention in the cryptopores of gibbsitic Acrustox (pores with diameter < 0.01 μm) due to the high energy (3500 kPa), which makes this water unavailable to the roots of coffee trees [48, 49] (**Figure 9**).

The authors mentioned in the previous paragraph point out that deep preparation and maintenance of Brachiaria sp. should be considered as the main factors of this management system. The additional surface applied gypsum (7 kg m^{-1}), act as the supporting factor in the structure of the soils. Carducci et al. [6], when evaluating the same soils in 3D images obtained by X-ray computed tomography, verified that kaolinitic Latosols presented high spatial variability of the soil structure. These pores resulted from the rapid and well-branched growth of the coffee root system [7, 8]. This is extremely relevant information given that the interactions between soil and root have been considered as a key element for the second green revolution aimed at maximizing production [50].

4. Water-use efficiency and plant responses

The water content in the soil profile is one of the main factors of growth and productive vigor of coffee, mainly because it is predominantly implanted in a dry land system. In this sense, the knowledge of soil water dynamics in the root zone in production areas is strategic because it predicts the success of agricultural activity. Management strategies can contribute to the efficient use of stored soil water from rainfall and enable positive responses to the crop.

In order to reduce the effects of water deficit, a plastic film (double-sided, black and white) was used as mulching covering the coffee growing line. Such management provided greater soil water storage up to 0.60 m in an Argisol (Ultisol), with soil moisture above 30% in the dry season, from May to September (**Figure 10**). In the topsoil, the soil moisture also remained higher, especially in warmer seasons, such as in January. These results coincide with the highest growth in stem height and diameter over the first year of coffee development [51, 52], showing that mulching may be an important alternative for keeping water in the root zone at the most critical time for crop development.

In a Cerrado Latosol cultivated with coffee under a conservation system [10], soil moisture was monitored daily during 2010 by means of a capacitance multi-sensor probe to a depth of 1.0 m [53, 54]. Throughout the evaluated period, the lowest moisture values were observed in the 0.50 to 0.75 m layer, indicating that the coffee tree extracted the largest amount of water at this depth (**Figure 11**), coinciding with significant presence of coffee roots [7] (**Figure 4**). In addition, in the months corresponding to the dry season in the region (June to August), it was observed low humidity values in the depth of 1.00 m, and thus deep water absorption, which may have contributed to reduce the water stress suffered by the plant. In this sense, the groove opening and limestone incorporation at 0.60 m associated with the application of additional gypsum may be important for the attenuation of water deficit.

An alternative for soil moisture monitoring is the use of remote sensors, given their repeatability characteristics, access to large areas and easy handling. However, it should be taken into account that coffee is a perennial crop with high root system activity at depth, and the use of remote sensor data to directly measure soil moisture is limited to a few centimeters below the surface (±5 cm) [55], not covering the entire area of water extraction by the roots [56]. Santos et al. [57] used the vegetation index EVI-2 to monitor the vegetative vigor of the coffee tree and to correlate it

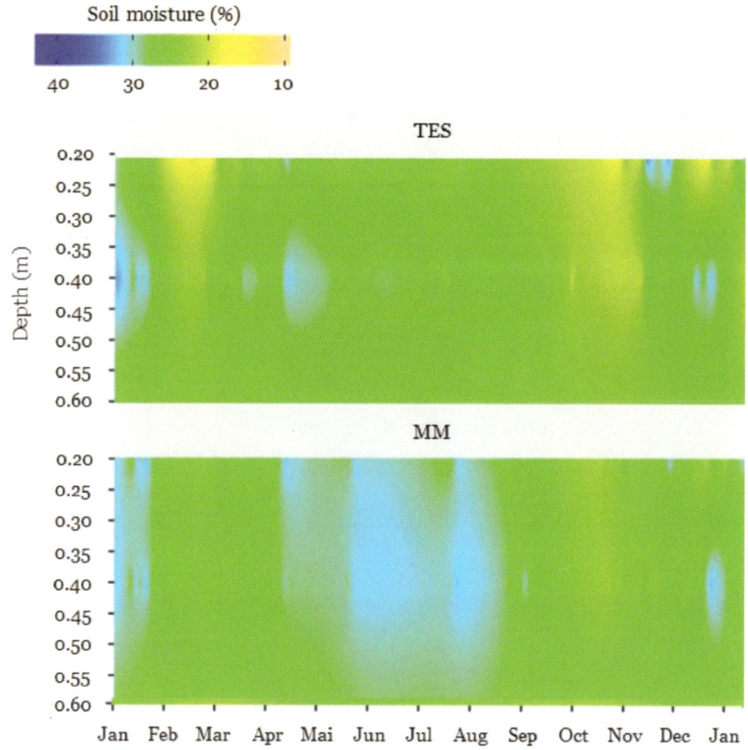

Figure 10.
Continuous variation of soil moisture in 2014 in the 0–0.60 m layer of Argisol as a function of conventional management (TES) and plastic cover (MM) in the first year of coffee cultivation. Source: Adapted from Barbosa [51].

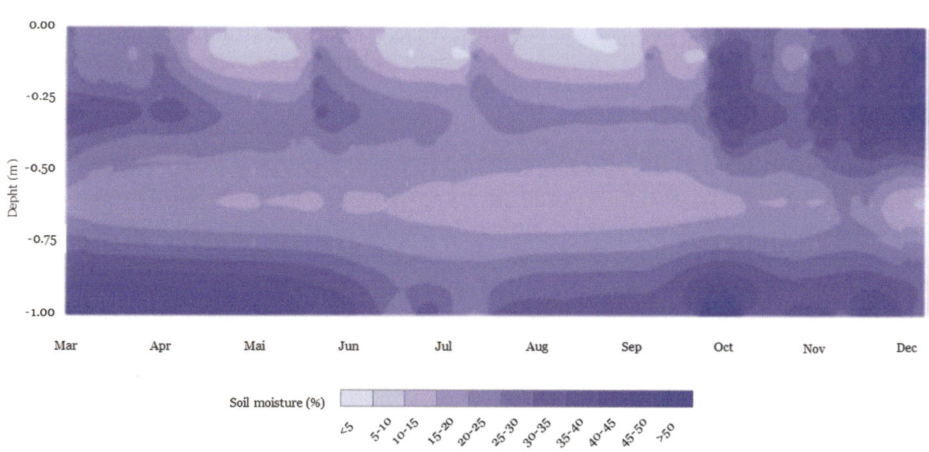

Figure 11.
Continuous variation of soil moisture (% by volume) in the planting line (sensor positioned 0.15 m from the coffee tree trunk) as a function of depth (0–1.00 m) and time (March/2010 to December/2010) in a very clayey gibbsitic-oxidic dystrophic red Latosol with coffee during the 2nd year under management system described in Serafim et al. [10], which includes deep preparation with chemical correction up to 0.60 m, cultivation of Brachiaria between the rows and application of additional gypsum at a dose of 28 Mg ha^{-1}. Source: Adapted from Silva [53].

with moisture data at different depths. The authors concluded that it is possible to estimate the water content in the root zone using EVI-2, and that the humidity at a depth of 0.60 m is the one that most reflects the water situation of the plant.

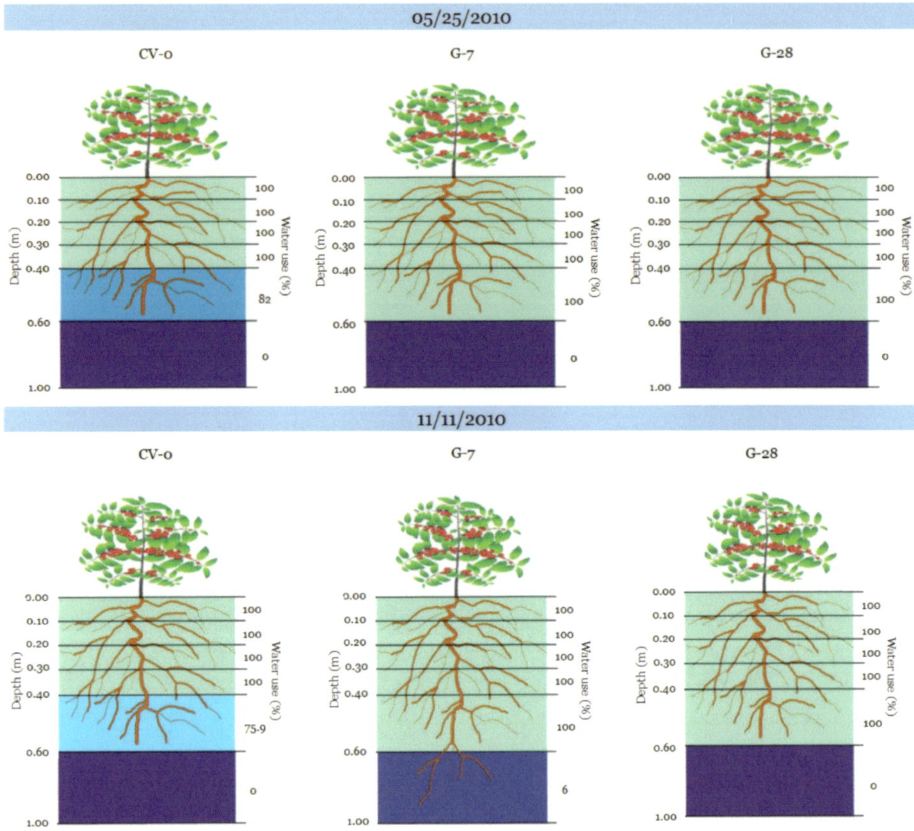

Figure 12.
Water use in layers up to 1.00 m depth during the dry season (May 2010) and in the summer (November 2010) for coffee trees installed in October/2008 due to management with deep soil preparation and limestone incorporation at 0.60 m depth, differing by presenting Brachiaria between the rows and additional application of 7 Mg ha^{-1} of gypsum (G-7) or 28 Mg ha^{-1} of gypsum (G-28), and without application of additional gypsum and uncovered line (CV-0). Source: Ivan Célio Andrade Ribeiro.

To detail the use of additional gypsum practice, water use by the coffee tree in the soil profile was estimated at different time intervals in 2010 (**Figure 12**). The coffee tree consumed water to a depth of 0.60 m in both evaluations performed and for all managements, which corroborates the lower moisture values in this layer (**Figure 11**), confirming the importance of deep tillage and soil correction at 0.60 m. The highest water consumption was observed for treatment G-7, followed by G-28 and lastly for CV-0. The use of additional gypsum allowed the development of thin roots in treatments G-7 and G-28 when compared with CV-0 [11], which may be due to the high levels of exchangeable Ca^{2+}, Mg^{2+} and K^{+} in the soil solution, which remained at adequate values to a depth of 0.85 m in the management with additional gypsum application [58].

Water use at a depth of 1.0 m was observed only in the G-7 treatment in November 2010, where the plant consumed about 6% of the stored water. At that time, the coffee tree was 2 years old, showing potential for deep water extraction. Moreover, even in the rainy season there was drought of more than 20 days [29, 53], associated with lower rainfall in the region this year compared to the historical average [59], implying less soil water storage. However, the high soil moisture at a depth of 1.00 m - above the critical moisture content for reducing maximum coffee perspiration in all managements [29] - indicates that this layer is an important water reservoir that can be accessed by plants during the driest or summer periods, reinforcing the importance of deepening the root system through management [29, 59].

Although management with additional gypsum (G-7 and G-28) provided higher water consumption compared to CV-0, it was not possible to differentiate its effects on water stress suffered by plants, evaluated by leaf water potential (ψf) in January, April, and August 2010 (**Figure 13**). It is noteworthy that in CV-0, although no additional gypsum was applied, liming was performed on the surface and in the planting furrow to a depth of 0.60 m, which favored the deepening of the root system [7].

The highest water stress was observed in August (**Figure 13**), coinciding with the peak of the dry season in the region and the lowest soil water content [29, 59]. However, all observed ψf values were below the critical range of water stress that leads to a reduction in coffee crop production, which is between −1.8 and −2.5 MPa [60–62].

Regarding plant growth, lower plant height values were observed in the G-7 and G-28 managements when compared with CV-0 (**Figure 13**), which may be explained by competition for root-shoot photoassimilates [63], since the coffee tree showed denser and deeper root systems for G-7 and G-28 [11]. In addition, the evaluations were carried out shortly after planting and, considering that the main morphological and physiological characteristics of the coffee root system complete its development at 5 years of age [1], it is expected that the investments made in liming, application additional gypsum and fertilization result in greater root development in the G-7 and G-28 managements in subsequent years [21].

Despite the lower initial plant growth, the adoption of the conservation management system provided maintenance of the water state of plants during the dry and summer season (**Figure 13**), resulting in statistically equal yields between CV-0, G-7 and G-28 management system at the first harvest in 2011 [29], highlighting the importance of deep tillage and soil correction. However, in 2012, higher yields were obtained in the managements G-7 and G-28. On average, production was 52.8 bags ha^{-1} in CV-0 (1 bag = 60 kg of coffee grains), 54.5 bags ha^{-1} in G-7 and 58.0 bags ha^{-1} in G-28 [59]. Coffee plants take 2 years to complete their phenological cycle [64]. Thus, soil moisture in 2010 influenced production in 2011 and 2012, demonstrating the positive effect of investments in additional gypsum associated with the

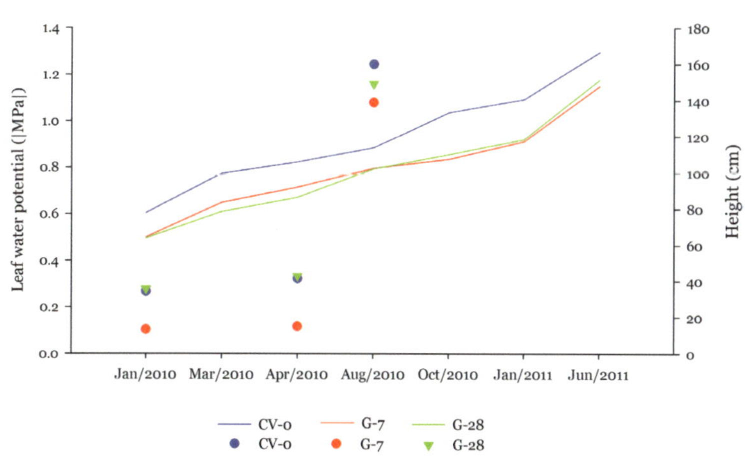

Figure 13.
Water stress assessed by leaf water potential (Ψf) at three times of the year (January 5; January 18 and August 20, 2010), and plant growth at height assessed continuously each month (January 5, 2010 to June 18, 2011) for coffee planted in October 2008 due to management with deep soil preparation and limestone incorporation, differing for presenting Brachiaria between the rows and additional application of 7 Mg ha^{-1} of gypsum (G-7) or 28 Mg ha^{-1} of gypsum (G-28), without application of additional gypsum and uncovered gypsum (CV-0). Source: From authors.

maintenance of Brachiaria in the G-7 and G-28 treatments, which provided higher water consumption by the plant in 2010 (**Figure 12**).

The trend of higher production for management with additional gypsum was confirmed in the 2013 crop, in which 63.0 bags ha^{-1} was produced in CV-0; 75.5 bags ha^{-1} in G-7, and 71.1 bags ha^{-1} in G-28 (data obtained through personal communication with consultants in the area). However, in the 2014 harvest, only the G-28 treatment presented higher yield (87.2 bags ha^{-1}) when compared with CV-0 (85.6 bags ha^{-1}). Management G-7 presented the lowest yield (57.5 bags ha^{-1}). However, when evaluating the general average of the first four seasons, it is observed that there is little difference between the evaluated managements, in which in CV-0 were harvested 63.6 bags ha^{-1}, 60.5 bags ha^{-1} in G-7 and 68.6 bags ha^{-1} in G-28.

5. Final remarks

There have been strong droughts and short-time droughts in rainy season in the main coffee producing regions of Brazil. Although most of the soils used are deep and capable of storing a large volume of water, these soils have a small effective depth for the development of the root system due to severe chemical limitations, therefore causing a yield gap.

Brazilian researchers have studied ways to overcome this problem, such as selecting drought tolerant plants. However, a strategy that has attracted the attention of farmers is the adoption of soil management systems that provide the best development of the coffee root system, with chemical and physical adequacy of soils. Deep tillage, maintenance of intercropped Brachiaria in the coffee plant interrow and additional gypsum play important roles in this management system. This is relevant information given that the interactions between soil and root have been considered as key elements for the maximization of crop production. Therefore, the set of practices previously mentioned in this chapter has alleviated the soil limitations caused by droughts, root growth and, consequently, the development of productive coffee trees.

Author details

Bruno Montoani Silva[1*], Geraldo César de Oliveira[1], Milson Evaldo Serafim[2], Carla Eloize Carducci[3], Érika Andressa da Silva[1], Samara Martins Barbosa[1], Laura Beatriz Batista de Melo[1], Walbert Junior Reis dos Santos[4], Thiago Henrique Pereira Reis[5], César Henrique Caputo de Oliveira[5] and Paulo Tácito Gontijo Guimarães[5]

1 Department of Soil Science, Federal University of Lavras, Brazil

2 Federal Institute of Education, Science and Technology of MatoGrosso, Brazil

3 Department of Agriculture, Federal University of Grande Dourados, Brazil

4 Federal Institute of Education, Science and Technology of Southern Minas Gerais, Brazil

5 Agricultural Research Company of Minas Gerais (Epamig), Brazil

*Address all correspondence to: brunom.silva@ufla.br

References

[1] Rena AB, Guimarães PTG. Sistema radicular do cafeeiro: estrutura, distribuição, atividade e fatoresque o influenciam. Belo Horizonte: EPAMIG; 2000

[2] Carducci CE, Oliveira GC, Zeviani WM, Lima VMP, Serafim ME. Bimodal pore distribution on soils under conservation management system for coffee crop. Engenharia Agrícola. 2013;**33**:291-302. DOI: 10.1590/S0100-69162013000200008

[3] Mazzafera P, Carvalho A. Produção e tolerância à seca de cafeeiros. Bragantia. 1987;**46**:403-415

[4] Carvalho FG, Sera GH, Andreazi E, Sera T, Fonseca ICB, Carducci FC, et al. Tolerância ao déficit hídrico em mudas de genótipos de café portadores de genes de diferentes espécies. Coffee Science. 2017;**12**:156-163

[5] Motta ACV, Nick JÁ, Yorinori GT, Monte SB. Distribuição horizontal e vertical da fertilidade do solo e das raízes de cafeeiro (*Coffea arabica* L.) cultivar Catuaí. Acta Scientiarum Agronomy. 2006;**28**:455-463. DOI: 10.4025/actasciagron.v28i4.758

[6] Carducci CE, Oliveira GC, Curi N, Rossoni DF, Costa AL, Heck RJ. Spatial variability of pores in oxidic latosol under a conservation management system with different gypsum doses. Ciência e Agrotecnologia. 2014;**38**:445-460. DOI: 10.1590/S1413-70542014000500004

[7] Carducci CE, Oliveira GC, de Lima JM, Rossoni DF, Costa AL, Oliveira LM. Distribuição espacial das raízes de cafeeiro e dos poros de dois latossolos sob manejo conservacionista. Revista Brasileira de Engenharia Agrícola e Ambiental. 2014;**18**:270-278. DOI: 10.1590/S1415-43662014000300005

[8] Silva EA, Silva SHG, Oliveira GC, Carducci CE. Root spatial distribution in coffee plants of different ages under conservation management system. African Journal of Agricultural Research. 2016;**11**:4970-4978. DOI: 10.5897/AJAR2016.11356

[9] Favarin JL, Almeida REM, Salustio PEB, Pedrosa AW. Novo conceito no preparo do solo para a lavoura cafeeira. Visão Agrícola. 2013;**12**:20-22

[10] Serafim ME, Oliveira GC, Oliveira AS, Lima JM, Guimarães PTG, Costa JC. Sistema conservacionista e de manejo intensivo do solo no cultivo de cafeeiros na região do alto São Francisco, MG: um estudo de caso. Bioscience Journal. 2011;**27**:964-977

[11] Carducci CE, Oliveira GCC, Curi N, Heck RJJ, Rossoni DFF, Carvalho TSS, et al. Gypsum effects on the spatial distribution of coffee roots and the pores system in oxidic Brazilian Latosol. Soil and Tillage Research. 2015;**145**:171-180. DOI: 10.1016/j.still.2014.09.015

[12] Silva EA, Carducci CE, Oliveira GC, Silva BM, Serafim ME. Estrutura de solos em manejo conservacionista: diagnóstico visual, laboratorial, caracterização e inter-relações. Scientia Agrária. 2017;**18**:61-73. DOI: 10.5380/rsa.v18i3.51646

[13] vanRaij B. Gesso na agricultura. Campinas: Instituto Agronômico; 2008. p. 233

[14] Conceição PC, Amado TJC, Mielniczuk J, Spa Gnollo E. Qualidade do solo em sistemas de manejo avaliada pela dinâmica da matéria orgânica e atributos relacionados. Revista Brasileira de Ciência do Solo. 2005;**29**:777-788. DOI: 10.1590/S0100-06832005000500013

[15] Bayer C, Martin-Neto L, Mielniczuk J, Pavinato A, Dieckow J. Carbon sequestration in two Brazilian Cerrado soils under no-till. Soil and Tillage Research. 2006;**86**:237-245. DOI: 10.1016/j.still.2005.02.023

[16] Resende M, Curi N, Rezende SB, Corrêa GF. Pedologia: base para distinção de ambientes. 6th ed. Lavras: Editora UFLA; 2014. p. 378

[17] Serafim ME, Oliveira GC, Curi N, Lima JM, Guimarães PTG, Lima VMP. Potencialidades e limitações de uso de Latossolos e Cambissolos, sob sistema conservacionista em lavouras cafeeiras. Bioscience Journal. 2013;**29**:1640-1652

[18] Serafim ME, Oliveira GC, Vitorino ACT, Silva BM, Carducci CE. Qualidade física e intervalo hídrico ótimo em Latossolo e Cambissolo, cultivados com cafeeiro, sob manejo conservacionista do solo. Revista Brasileira de Ciência do Solo. 2013;**37**:733-742. DOI: 10.1590/S0100-06832013000300020

[19] Serafim ME, Oliveira GC, Lima JM, Silva BM, Zeviani WM, Lima VMP. Disponibilidade hídrica e distinção de ambientes para cultivo de cafeeiros. Revista Brasileira de Engenharia Agrícola e Ambiental. 2013;**17**:362-370. DOI: 10.1590/S1415-43662013000400002

[20] Silva EA. Propriedades físico-hídricas do solo e desenvolvimento radicular do cafeeiro [dissertation]. Lavras: Universidade Federal de Lavras; 2017

[21] Silva BM, Oliveira GC, Serafim ME, Silva A, Ferreira MM, Norton LD, et al. Critical soil moisture range for a coffee crop in an oxidic latosol as affected by soil management. Soil and Tillage Research. 2015;**154**:103-113. DOI: 10.1016/j.still.2015.06.013

[22] Busscher WJ, Bauer PJ, Frederick JR. Recompaction of a coastal loamy sand after deep tillage as a function of subsequent cumulative rainfall. Soil and Tillage Research. 2002;**68**:49-57. DOI: 10.1016/S0167-1987(02)00083-1

[23] Busscher WJ, Edwards JH, Vepraskas MJ, karlen DL. Residual effects of slit tillage and subsoiling in a hardpan soil. Soil and Tillage Research. 1995;**35**:115-123. DOI: 10.1016/0167-1987(95)00488-2

[24] Marcolan AL, Anghinoni I. Atributosfísicos de um Argissolo e rendimento de culturas de acordo com o revolvimento do solo em plantio direto. Revista Brasileira de Ciência do Solo. 2006;**30**:163-170. DOI: 10.1590/S0100-06832006000100016

[25] Benevenute PAN. Efeitos do preparo do sulco de plantio nos atributos físico-hídricos do solo, no desenvolvimento radicular de plantas e em parâmetros de produção do citros [dissertation]. Lavras: Universidade Federal de Lavras; 2019

[26] AMS J, Carvalho SP, Soares ÂM. Comparação entre sistemas radiculares de mudas de *Coffea arabica* L. obtidas por estaquia e por sementes. Coffee Science. 2006;**1**:14-20

[27] Ferreira MM, Fernades B, Curi N. Mineralogia da fração argila e estrutura de latossolos da região sudeste do Brasil. Revista Brasileira de Ciência do Solo. 1999;**23**:507-514. DOI: 10.1590/S0100-06831999000300003

[28] Carducci CE, Oliveira GC, Curi N, Heck RJ, Rossoni DF. Scaling of pores in 3D images of latosols (oxisols) with contrasting mineralogy under a conservation management system. Soil Resource. 2014;**52**:231-243. DOI: 10.1071/SR13238

[29] Silva BM, Oliveira GC, Serafim ME, Silva ÉA, Guimarães PTC, Melo LBB,

et al. Soil moisture associated with least limiting water range, leaf water potential, initial growth and yield of coffee as affected by soil management system. Soil and Tillage Research. 2019;**189**:36-43. DOI: 10.1016/j.still.2018.12.016

[30] Silva CA, Guimarães PTG. Cafeeiro: vetores para o aumento da produtividade. In: Casarin V, editor. Informações Agronômicas número 155. Piracicaba: IPNI; 2016. pp. 13-16

[31] Resende AV, Fontoura SM, Borghi E, Santos FC, Kappes C, Moreira SG, et al. Solos de fertilidade construída: características, funcionamento e manejo. In: Casarin V, editor. Informações Agronômicas número 156. Piracicaba: IPNI; 2016. pp. 1-19

[32] Nunes UR, Andrade Júnior VC, Silva EB, Santos NF, Hao C, Ferreira CA. Produção de palhada de plantas de cobertura e rendimento do feijão em plantio direto. Pesquisa Agropecuária Brasileira. 2006;**41**:943-948. DOI: 10.1590/S0100-204X2006000600007

[33] Souza LS, Domingues MCS, Bertoncini GA. Efeito da incorporação do capim braquiaria no solo sobre o desenvolvimento do milho com diferentes fontes de nitrogênio. In: Unimar Ciências. 2013;**1-2**:17-22

[34] Guimarães PTG, Dias KGL, Oliveira CHC. Uso de Braquiária nas entrelinhas do cafeeiro. Circular Técnica Epamig; 2018

[35] Guimarães PTG, Reis THP. Nutrição e Adubação do Cafeeiro. In: Reis PR, Cunha RL, editors. Café Arábica: do plantio à colheita. Belo Horizonte: Epamig SM; 2010. pp. 343-414

[36] Revista Cafeicultura. Raiz de Brachiaria Ruzizienses atinge cinco metros de profundidade [Internet]. 2018. Available from: https://www.youtube.com/channel/UCR9RPOgh0nzVhrG52skZJFw. [Accessed: 19 August 2019]

[37] Crispim SMA, Branco OD. Aspectos gerais das Braquiárias e suas características na sub-região da Nhecolândia, Pantanal, MS. 1st ed. Corumbá: Embrapa Pantanal; 2002. p. 25

[38] Ribeiro AC, Guimarães PTG, Alvarez VVH. editors. Recomendações para o uso de corretivos e fertilizantes em Minas Gerais - 5ª Aproximação, Viçosa: SBCS; 1999. p. 359

[39] Alcântara EM, Cunha RL, Silva RA. Manejo do mato em cafeeiros: métodos e coeficientes técnicos utilizados. Informe Agropecuário. 2008;**29**:74-82

[40] DaMatta FM, Ramalho JDC. Impacts of drought and temperature stress on coffee physiology and production: A review. Brazilian Journal of Plant Physiology. 2006;**18**:55-81. DOI: 10.1590/S1677-04202006000100006

[41] Hillel D. Solo e água: fenômenos e princípios físicos. Porto Alegre: Universidade Federal do Rio Grande do Sul; 1970. p. 231

[42] Rezende JO. Compactação e adensamento do solo, metodologia para avaliação e práticas agrícolas recomendadas. In: XXVI Congresso Brasileiro de Ciência do Solo; 1997; Rio de Janeiro. Rio de Janeiro: Sociedade Brasileira de Ciência do Solo; 1997. CD-ROM

[43] Klein VA, Libardi PL. Densidade e distribuição do diâmetro dos poros de um Latossolo Vermelho sob diferentes sistemas de uso e manejo. Revista Brasileira de Ciência do Solo. 2002;**26**:857-867. DOI: 10.1590/S0100-06832002000400003

[44] Volland-Tuduri N, Bruand A, Brossard M, Balbino LC, Oliveira MIL, Martins ÉS. Mass proportion of

microaggregates and bulk density in a Brazilian clayey oxisol. Soil Science Society of America Journal. 2005;**69**: 1559-1564. DOI: 10.2136/sssaj2003.0344

[45] Cooper M, Vidal-Torrado P. Caracterização morfológica, micromorfológica e físico-hídrica de solos com horizonte B Nítico. Revista Brasileira de Ciência do Solo. 2005;**29**:581-595. DOI: 10.1590/S0100-06832005000400011

[46] Oliveira GC, Dias Júnior MS, Resck DVS, Curi N. Caracterização química e físico-hídrica de um Latossolo Vermelho após vinte anos de manejo e cultivo do solo. Revista Brasileira de Ciência do Solo. 2004;**28**:327-336. DOI: 10.1590/S0100-06832004000200011

[47] Carducci CE, Oliveira GC, Curi N, Severiano EC, Zeviani WM. Relations of clay fraction mineralogy, structure and water retention in oxidic latosols (oxisols) from the Brazilian Cerrado biome. In: Valaskova M, Martynková GS, editors. Clay Minerals in Nature. 1st ed. Croatia: Intech Open; 2012. pp. 149-170. DOI: 10.5772/47785

[48] Carducci CE, Oliveira GC, Severiano EC, Zeviani WM. Modelagem da curva de retenção de água de latossolos utilizando a equação duplo van Genuchten. Revista Brasileira de Ciência do Solo. 2011;**35**:77-86. DOI: 10.1590/S0100-06832011000100007

[49] Severiano EC, Oliveira GC, Dias Júnior MS, Curi N, Costa KAP, Carducci CE. Preconsolidation pressure, soil water retention characteristics, and texture of latosols in the Brazilian Cerrado. Soil Resource. 2013;**51**:193-202. DOI: 10.1071/SR12366

[50] Lynch JP. Roots of the second green revolution. Australian Journal of Botany. 2007;**55**:493-512. DOI: 10.1071/BT06118

[51] Barbosa SM. Condicionamento físico-hídrico do solo como potencializador do crescimento inicial do cafeeiro [thesis]. Lavras: Universidade Federal de Lavras; 2015

[52] Barbosa SM. Manejo de Argissolos e Cambissolos na implantação de cafeeiros [dissertation]. Lavras: Universidade Federal de Lavras; 2018

[53] Silva BM. Dinâmica espaço-temporal da água no solo cultivado com cafeeiro nas condições climáticas do Alto São Francisco – MG [thesis]. Lavras: Universidade Federal de Lavras; 2012

[54] Silva BM, Santos WJR, Oliveira GC, Lima JM, Curi N, Marques JJ. Soil moisture space-time analysis to support improved crop management. Ciência e Agrotecnologia. 2015a;**39**:39-47. DOI: 10.1590/S1413-70542015000100005

[55] Crosson WL, Limaye AS, Laymon CA. Parameter sensitivity of soil moisture retrievals from airborne C and X-band radiometer measurements in SMEX02. IEEE Transactions on Geoscience and Remote Sensing. 2005;**43**:2842-2853

[56] Liu S, Roberts DA, Chadwick OA, Still CJ. Spectral responses to plant available soil moisture in a Californian grassland. International Journal of Applied Earth Observation and Geoinformation. 2012;**9**:31-44. DOI: 10.1016/j.jag.2012.04.008

[57] Santos WJR, Silva BM, Oliveira GC, Volpato MML, Lima JM, Curi N, et al. Soil moisture in the root zone and its relation to plant vigor assessed by remote sensing at management scale. Geoderma. 2014;**221-222**:91-95. DOI: 10.1016/j.geoderma.2014.01.006

[58] Ramos BZ, Toledo JPVF, Lima JM, Serafim ME, Bastos ARR, Guimarães PTG, et al. Doses de gesso em cafeeiro: influência nos teores de cálcio, magnésio, potássio e pH na solução de um Latossolo Vermelho distrófico. Revista Brasileira de Ciência

do Solo. 2013;**37**:1018-1026. DOI: 10.1590/S0100-06832013000400019

[59] Silva BM. Disponibilidade de água no solo: métodos de estimativa e implicações de manejo em cafeeiros na região do cerrado [dissertation]. Lavras: Universidade Federal de Lavras; 2014

[60] Guerra AF, Rocha OC, Rodrigues GC. Manejo do cafeeiro irrigado no Cerrado com estresse hídrico controlado. Irrigação e Tecnologia Moderna. 2005;**65-66**:45-45

[61] Silva AL, Bruno IB, Reichardt K, Bacchi OOS, Dourado-Neto D, Favarin JL, et al. Soil water extraction by roots and Kc for the coffee crop. Revista Brasileira de Engenharia Agrícola e Ambiental. 2009;**19**:257-261. DOI: 10.1590/S1415-43662009000300006

[62] Soares AR, Mantovani EC, Braga A. Irrigação e fisiologia da floração em cafeeiros adultos na região da zona da mata de Minas Gerais. Acta Scientiarum. Biological Sciences. 2005;**27**:117-125. DOI: 10.4025/actasciagron.v27i1.2128

[63] DaMatta FM, Rochi CP, Maestri M, Barros RS. Ecophysiology of coffee growth and production. Brazilian Journal of Plant Physiology. 2007;**19**:485-510. DOI: 10.1590/S1677-04202007000400014

[64] Camargo ÂP, Camargo MBP. Definição e esquematização das fases fenológicas do cafeeiro arábica nas condições tropicais do Brasil. Bragantia. 2001;**60**:65-68. DOI: 10.1590/S0006-87052001000100008

Chapter 3

Soil Electrochemical and Physical Properties in Coffee Crops in the State of Paraná, Brazil

Cezar Francisco Araujo-Junior, Vinicius Cesar Sambatti, João Henrique Vieira de Almeida Junior and Henrique Hiroki Yamada

Abstract

In aerated soils, pH is considered to control available nutrients to plants. Also, pH is related to soil charges and is a key property expanding double layers of colloids. These electrochemical properties are affected by soil management and are related to soil physical properties like water-dispersible clay, aggregation indexes, and infiltration rate. Water-dispersible colloids are the fraction of clay that disperse in water and are affected by nature of soil including mineralogy of clay fraction, soil management in terms of crop sequence, application of organic manures, field traffic, and mechanical stress by time of shaking for the analysis. Traffic of machines, soil tillage, and weed control methods are the main causes of change in soil physical properties in coffee crop. However, management of soil acidity with limestone and use of gypsum also can change soil physical and electrochemical properties, which are related with dynamic processes like soil air and hydraulic permeability into soil which are essential to root development and growth. Therefore, soil management in coffee crops requires comprehension of the effects of changes in soils caused by addition of amendments like limestone and gypsum, traffic of machines, and weed control methods on behavior of soil properties for better management.

Keywords: *Coffea arabica* L., point of zero charge (PZC), water-dispersible clay, soil-water retention, soil structure, weed management

1. Introduction

Crop production is more problematic in tropical and subtropical climatic conditions than in temperate humid climates [1]. These authors highlighted that many cultivated plants derive from regions which are ecologically different from their present region of production. When competing with weeds which have become properly adapted to their habitat, these crops cannot survive without protection and assistance from the farmer.

Brazil is the greatest exporter of coffee and second consumer of the product in the world. Currently, the land area cultivated with coffee is 1.84 million of hectares, the total 1.47 million of hectare (≈80%) being cultivated with *Coffea arabica* L. species and 373.33 mil hectares (≈20%) with *Coffea canephora* Pierre (Secretary of Agricultural Policy in the Ministry of Agriculture, Livestock and Food Supply—MAPA, 2019).

In the state of Paraná, most coffee plants are cultivated in high coffee population density system (>5000 plants per hectare) which guarantees the coffee farmers higher productivity by hectare [2] in soils derived from basalt with high iron content (Fe_2O_3 higher than 18 dag kg^{-1}). In these soils, the workability is easier, and the drainage is usually very good [3]. On the other hand, in the high coffee population density, spacing between coffee rows is 2–3.2 m and 0.5–0.75 m between the plants in the lines, which compromises the mechanization of operations to management and harvest.

Soil slope is another important factor that compromises the mechanical operations and coffee management besides coffee planting spacing. In a previous study done in the climatic zoning of the State of Paraná for the cultivation of coffee (**Figure 1**), Höfig and Araujo-Junior [3] showed that 89% of the land area are not limited to mechanization by the criterion of soil slope classes.

Considering mesoregions of the State of Paraná, the northwest has the smallest area with a slope higher than 20% and therefore the greatest potential for mechanization of coffee plantations. On the other hand, in the Pioneer

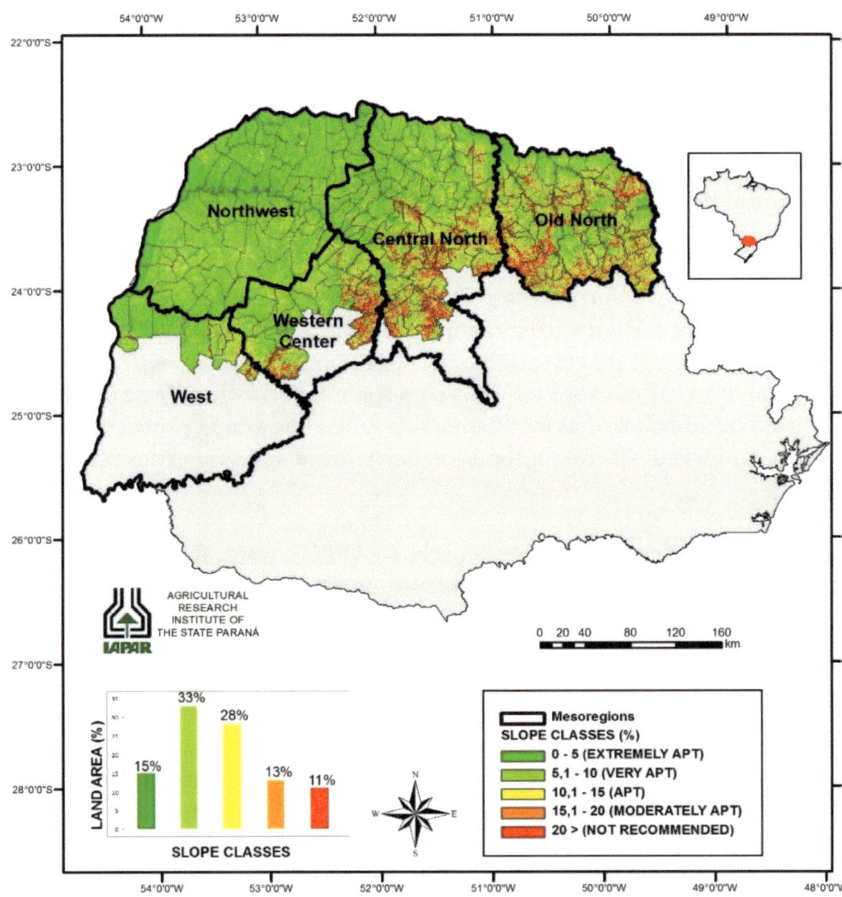

Figure 1.
Map of the State of Paraná with slope classes and potential for mechanization in the area with climatic zoning to cultivate coffee. Source: Höfig and Araujo-Junior [3].

Northern mesoregion, which currently has about 37% of coffee plantations in the State of Paraná and has great technological potential due to edaphoclimatic characteristics for coffee cultivation, 10% of the area is not recommended for mechanization based on soil slope classes, which represents an area of 160,000 hectares.

Besides mechanization, the addition of soil amendments like limestone and gypsum and weed control methods affects the behavior of soil properties. In long-term experiments conducted in different coffee regions of Brazil, the effects of weed control methods on soil attributes have been proven.

Weeds when properly managed in both row and interrow areas can become allied with the coffee farmer without compromising crop yield. On the other hand, when the weed is constantly controlled with pre-emergence herbicide, soil surface is exposed to direct raindrop impact which can form soil crusting that makes water infiltration more difficult. As a consequence of surface soil crusting, runoff is increased, and hydric soil erosion must be a problem.

Research results have shown that weed management modifies soil resistance to compaction and can minimize damage caused by machine traffic on the soil as well as assisting soil and water management and conservation by providing benefits for accelerated water erosion.

In this chapter of the book, we presented that soil physical and mechanical properties are essential to the assessment of the effect of anthropogenic activities on natural resources and may help coffee farmers to obtain an optimum soil environment for plant growth. Due to that, this chapter characterizes the soil physical properties in coffee crops in the State of Paraná, Southern Brazil.

2. Soil electrochemical and physical properties in coffee crops

Soil physical properties are essential to comprehension of the behavior of the soil when submitted to mechanization. Among the physical properties, the water content in the soil profile determines the reaction to tillage, and soil moisture is the most important for soil-machine interactions, since it controls the consistency of the soil [4] and governs the amount of soil deformation when subjected to external pressure [5].

In aerated soils, pH is considered to control available nutrients to plants. Also, pH is related to soil charges and is a key property expanding double layers of colloids. Points of zero charge (PZC) are pH values associated with specific conditions imposed on one or more surface charge densities of an electrified interfacial region between a soil solution and soil solid phase [6]. Point of zero charge indicates the pH at which the net surface charge on variable charge surface is zero [7].

The surface charge density is the most important physical characteristic of an electrified interface. It can be defined as the number of coulombs per square meter borne by surface functional groups, and it depends in sign and magnitude on the composition of the soil solution and structure of the solid phase to which the functional groups are bound [6].

PZC was estimated from pH-H_2O and pH-KCl (1 mol L^{-1}) through Eq. (1). According to these authors, ΔpH is the electrochemical soil properties that most affect the dispersion. When ΔpH tends to zero, charges are balanced with less dispersion.

$$PZC = 2\,pH_{KCl} - pH_{H2O} \quad (1)$$

where PZC is the point of zero charge, dimensionless; pH_{kCl} is the hydrogenionic potential determined in potassium chloride solution (1 mol L^{-1}); pH_{H2O} is the hydrogenionic potential determined in water.

$$\Delta pH = pH_{KCl} - pH_{H2O} \qquad (2)$$

These electrochemical properties are affected by soil management and are related to soil physical properties like water-dispersible clay, aggregation indexes, and infiltration rate.

Water-dispersible colloid (WDC) is generally recognized as the fraction of clay that disperses in water. Dispersion is the ultimate state of breakdown that results in release of clay particles as a consequence of expanding double layers and dominating repulsive forces [8]. WDC is affected by the nature of soil including mineralogy [9–11], clay content and application of sewage sludges [7], soil management [8] in terms of crop sequence, application of organic manures [12, 13], soil tillage, and traffic [14] which have been shown to affect dispersion-flocculation of clay in soil structural elements.

Besides liming and gypsum on coffee crop, soil organic matter may have a dispersive or aggregating effect according to the quantity and quality of the fertilizer [12]. These authors observed that the addition of manure at the doses of 23 g kg^{-1} and 30 g kg^{-1} provided dispersion of the clay fraction in electropositive *red-yellow* Latosol caused by negative electric charge balance.

3. Mechanization in crops with high coffee shrub population density

In coffee plantations, mechanization has emerged as an alternative in reducing production costs, operating income, and reducing labor hardness. However, for the preservation of natural resources (soil and water), soil and machinery management becomes essential to minimize the effects of anthropogenic actions on the soil. Within these aspects the coffee farmers can reduce the axle load and the contact pressure of the tires with the soil and use management systems that contribute to the deposition of organic matter, as soils with greater aggregation support more load and the organic matter relieves stress exercised by agricultural machinery.

Soil stress exerted by tires or tracks of machines on soil interface can be assessed according to machine characteristics and soil attributes [15]. These authors modelled soil stress through the software Tyres/Tracks and Soil Compaction (TASC) version 3.0 [16].

TASC version 1.0 was used for the first time in Brazil in a long-term weed control method experiment to assess the effects of weed control on soil load-bearing capacity and the impact of a coffee tractor on soil stresses [17]. In this study, a coffee tractor Valmet®, model 68, with a power rating of 44.9 kW (61 hp), a total weight of 38,245 N (3900 kg), front tires of 6.16 at inflation pressure 172 kPa, wheel mass of 683 kg and rear tires of 12.4 R28 at inflation pressure 124 kPa, and wheel mass of 1.365 kg was used for coffee management and mechanical weed control.

The contact area at soil-tire interfaces ranged from 0.0381 m^2 for the front tire to 0.1328 m^2 for the rear tire 12.4 R28 [17]. The ground average contact pressure at soil-tire interface ranged from 101 kPa to 176 kPa, with the highest occurring for the front tires. As highlighted by Guimarães Júnnyor et al. [15], the ground average contact pressure depends on the tire type, tire structure, tire sizes, wheel load, inflation pressure, and soil stiffness.

The State of Paraná (**Figure 1**) is located between the coordinates 22°30′ and 26°30′ S latitude and 48° and 55°W longitude within a region of climatic transition, from north to south, with regional and local variation due to altitude and topography. However, the recommended area to coffee cultivation in the State of Paraná [18] is located between the coordinates of the southern latitudes 22°51′50″, 24°76′70″ west longitude, and 49°55′30″ to 54°34′20″ from Greenwich (**Figure 1**).

High coffee shrub population density system (higher than 5000 shrubs per hectare) was developed in the State of Paraná by the Agronomic Institute of Paraná (IAPAR) to improve coffee yields and to provide economic benefits to coffee farmers; soil and water conservation also helps coffee shrubs after frost damage.

In Londrina in the State of Paraná, in an experiment conducted between 1976 and 1981 to determine the effects of spacing between plants in the row (4 m × 1 m and 4 m × 2 m), Siqueira et al. [19] concluded that irrespective of coffee cultivars (Catuaí Vermelho LCH 2077-2-5-81 and Acaiá LCP 474-4) or hybrid Icatu H 4782-7 AMBR (*Coffee arabica* vs *Coffea canephora* Pierre) the coffee yields per area increased with decreases of spacing between plants 2 m and 1 m [19].

In addition to crop response, coffee shrub population density system promotes improvements in chemical, physical, and biological soil attributes [20].

4. Liming and gypsum on clay fraction flocculation and soil particle aggregation

Soil correction by liming and/or gypsum has a significant influence on the soil physical and water properties [9, 21–26].

Liming by applying limestone–calcium carbonate [$Ca(CO_3)_2$] or magnesium carbonate [$Mg(CO_3)_2$] is the soil management practice used to correct excessive soil acidity. In addition to correcting acidity, lime application in soils is able to provide calcium and magnesium, provide nutrients, and neutralize excess aluminum and soil manganese that are toxic to plants [20, 22].

Gypsum applied on the surface of soil columns with dimensions of 0.6 m in height by 0.3 m in diameter provided increases in Ca contents and decrease in exchangeable Al contents, consequently favoring the root growth of deep coffee seedlings [27]. These authors also pointed out that the superficial application of the gypsum–$CaSO_4$–was more efficient than $CaCO_3$ incorporated at 0.3 m depth due to the higher root growth in depth as a result of the exchangeable calcium increase and aluminum reduction in the subsurface ground.

In the long term, both liming and gypsum can contribute to reducing the risks of erosion in coffee-cultivated LVdf, especially under conditions without green cover between rows and with uncovered soil. In a typical Distroferric Red Latosol (LVdf) (Rhodic Hapludox), very clayey texture cultivated with coffee shrubs in Londrina, Northern State of Paraná, liming and gypsum had positive effects on soil aggregation and consequently on the water infiltration rate in the soil profile 2 years after corrective application [21].

On the other hand, the short-term incubation studies (3 months after limestone application), using samples from an LVdf from Londrina, Castro Filho and Logan [9], found that the aggregates were stable in water up to the pH value in $CaCl_2$ equal to 5.7. On the other hand, when pH values exceeded 5.7, they reduced the stability of the aggregates in water.

5. Electrochemical properties of an Oxisol cultivated with coffee crop and its relationship with flocculation-dispersion of colloids and aggregates

5.1 Water-dispersible clay

Water-dispersible clay is the fraction of clay that disperses in water, and dispersive soils are common problematic soils in many parts of the world [14]. Due to fact that dispersible clay blocked soil porous system and compromise the water and gas movement, this soil property has been used in studies of hydric erosion and soil management.

In a long-term experiment conducted at experimental station of the IAPAR in Londrina, PR submitted to seven different weed management in the interrow area (between coffee row), weed managements in the interrow area did not affect the point of zero charge. The experiment was installed in randomized complete block design, seven treatments with four replicates. The soil in the experiment site is classified as Typical Dystroferric Red Latosol, very clayey texture (80 dag kg^{-1} clay) with kaolinitic mineralogy derived from the saprolites from basaltic rocks. The weed management were as follows: T1, hand-hoe weeding (HAWE); T2, portable mechanical mower (PMOW); T3, herbicides (HERB); T4, cover crop peanut horse (CCPH); T5, cover crop dwarf mucuna (CCDM); T6, no-weeding control in the interrow area (SCAP); and T7, weed check (CHECK) (no-weed control in the row and interrow area). Soil samples were collected at 0–10 cm, 10–20 cm, 20–30 cm, and 30–40 cm depths. Soil electrochemical properties were determined: pH in relation 1:2.5 soil/solution in a 0.01 mol L^{-1} $CaCl_2$, KCl, and water and point of zero charge (PZC). Water-dispersible clay was determined by pipette method without chemical dispersion shaker for 2 h.

However, the estimated point of zero charge changed among the soil sample depths (**Figure 2**). The estimated PCZ values were 4.2 (StDv. 0.54), 3.6 (0.54), 3.7 (0.29), and 3.7 (0.37) lower than the pH in all depths which contribute to excess in negative charge. When soil pH is higher than PZC, there were electrostatic repulsion and drop in clay flocculation [7].

For soil samples collected at county Londrina, Northern Paraná in Native Forest at 0–20 cm depth of a Red Latosol with very clayey texture (72 dag kg^{-1}), incubated with limed sludge with doses 1.5–24 g kg^{-1}, Tavares Filho et al. [7] observed after 180 days incubation in pots in the greenhouse PZC 4.8–5.03; delta pH −0.22 to −0.15; and water-dispersible clay 66–128 dag kg^{-1}; soil samples were shacked in an orbital shaker at 300 rpm for 3 h.

Weed management in the interrow area of coffee crop changes delta pH ($p < 0.001$) for a Dystroferric Red Latosol, very clayey at 0–10 cm depth. The highest values were found for T6 no-weed control between coffee rows ($\Delta pH = -0.57$) = T1 hand weeding ($\Delta pH = -0.68$) = T7 weed check ($\Delta pH = -0.72$) and lowest for T2 portable mechanical mower ($\Delta pH = -0.80$) = T3 herbicides ($\Delta pH = -0.93$) = T4 cover crop peanut horse ($\Delta pH = -0.80$) = T5 cover crop dwarf mucuna ($\Delta pH = -0.78$).

Changes in ΔpH affected the water-dispersible clay (**Figure 3**).

With the aim of to investigate the effect of liming on chemical and physical properties of three very fine, ferruginous, isothermic Rhodic Hapludox with different levels of organic carbon, Castro Filho [6] observed that pH affects positively the aggregate stability indexes. This author also suggested that aggregate stability depends on the soil mineralogical composition and the highest aggregation occurred nearly 100% of Al neutralization.

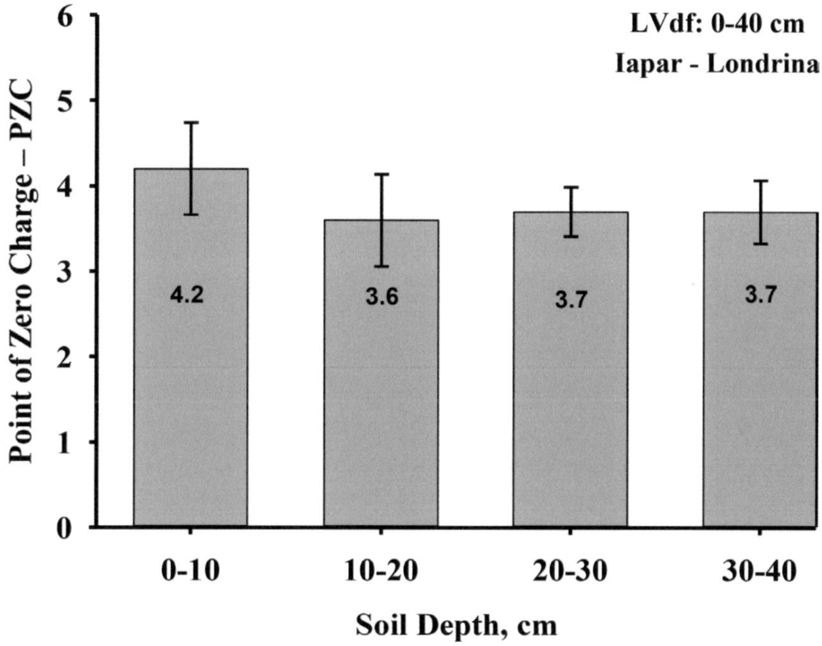

Figure 2.
Point of zero charge (PZC) in four layers of a Red Latosol cultivate coffee.

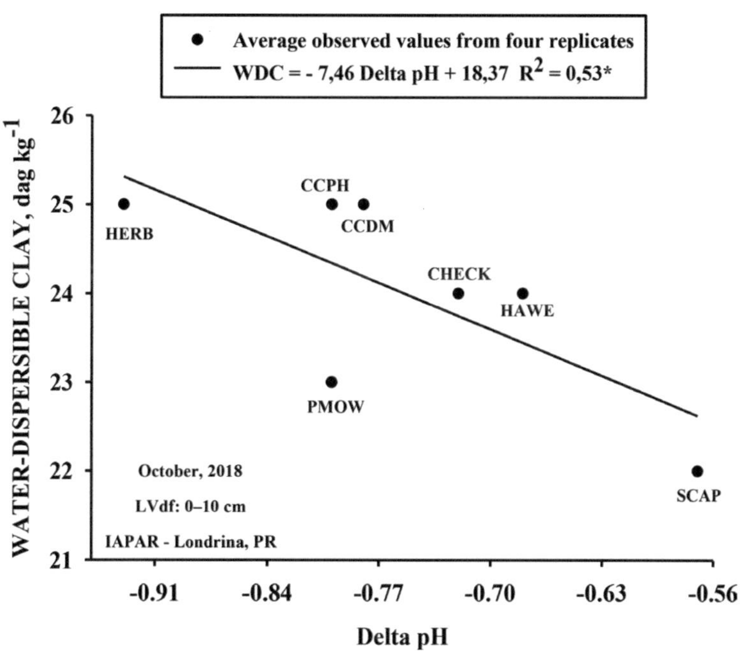

Figure 3.
Water-dispersible clay in relation to delta pH of a Dystroferric Red Latosol cultivated with coffee crop submitted to weed management in the interrow area. HAWE, hand weeding; PMOW, portable mechanical mower; HERB, herbicides; CCPH, cover crop peanut horse; CCDM, cover crop dwarf mucuna; SCAP, no-weed between coffee rows; CHECK, no-weeding in the interrow and row areas.

As can be observed in **Figure 4**, mean weight diameter increased as soil pH. Mean weight diameter (MWD) is large if the soil has a high percentage of large aggregates [6].

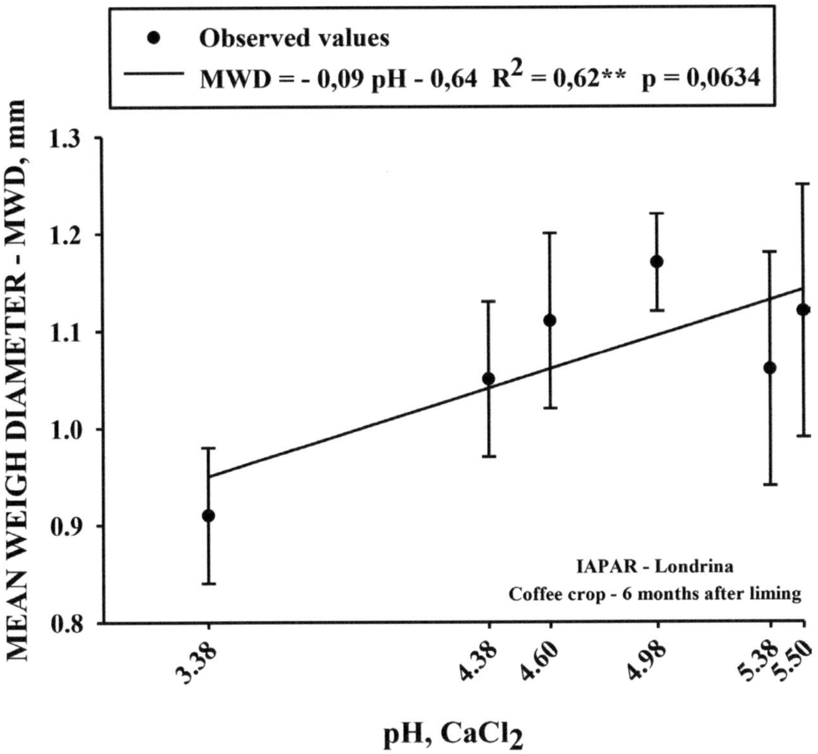

Figure 4.
Mean weight diameter (MWD) of the aggregates of a Red Latosol cultivate coffee 6 months after liming. Source: From Roth et al. [21].

Soil organic matter and soil pH (**Figure 3**), whose mechanisms involved depend on the substitution of aluminum by calcium in the sortive complex, participate in the soil aggregation whose formation and stabilization of the different classes of soil aggregate sizes will allow more or less lower aggregation, resulting in greater or lesser soil loss [9].

In a field experiment conducted in a Rhodic Hapludox cultivated with coffee, 2 years after the surface liming, Roth et al. [21] highlighted that the aggregation of solid particles exerts a significant action on soil susceptibility to accelerated water erosion for uncovered soil conditions. This study showed that after 60 min of simulated rainfall at an intensity of 60 mm per hour, soil maintained without soil correction with pH = 5.2 provided total infiltration of 56% of the total precipitation [21]. On the other hand, the authors observed that the best liming treatment to increase pH 7.0 provided 83% of total infiltration. In soil with pH 6.0 and with the application of plaster, the total infiltration was 67% of the total precipitation.

6. Conclusions

Traffic of machines, soil tillage, and weed control methods are the main causes of change in soil physical properties in coffee crop. However, management of soil acidity with limestone and use of gypsum also can change soil physical and electrochemical properties, which are related with dynamic processes like soil air and hydraulic permeability into soil which are essential to root development and growth.

Weed management in the interrow area of coffee crop changes delta pH for a Dystroferric Red Latosol very clayey at 0–10 cm depth. Changes in ΔpH affected the water-dispersible clay.

Acknowledgements

The authors are grateful to the IAPAR and Brazilian Consortium for Coffee Research and Development (CBP&D—Café) for providing financial support for this study. Also, this study was financed in part by the Coordination for the Improvement of Higher Education Personnel (CAPES)—Finance Code 001.

Conflict of interest

The authors declare that there is no conflict of interest anyway including that they are not members of the academic editors in Intech and members of an organization that could benefit financially or materially from the publication of their work.

Author details

Cezar Francisco Araujo-Junior[1,2]*, Vinicius Cesar Sambatti[3], João Henrique Vieira de Almeida Junior[4] and Henrique Hiroki Yamada[5]

1 Agronomic Institute of the State of Paraná – IAPAR, Soils Area, Brazil

2 Post Graduate Programme in Conservation Agriculture, Londrina, Brazil

3 Post Graduate Programme in Conservation Agriculture with Support Scholarship Offered by Coordination for the Improvement of Higher Education Personnel (CAPES)—Finance Code 001, Londrina, Brazil

4 Agronomy at State University of Londrina—UEL, Scientific Initiation Programme at IAPAR—ProICI with Support Scholarship Offered by National Council for Scientific and Technological Development (CNPq), Brazil

5 Agronomy at Center University of Filadélfia—UNIFIL, Scientific Initiation Programme at IAPAR—ProICI with Support Scholarship Offered by National Council for Scientific and Technological Development (CNPq), Brazil

*Address all correspondence to: cezar_araujo@iapar.br

References

[1] Krause R, Lorenz F. Soil Tillage in the Tropics and Subtropics. Eschborn: Deutsche Gesellschaft fur Technische Zusammenarbei (GTZ); 1984. p. 320

[2] Agronomic Institute of the State of Paraná—IAPAR. Modelo tecnológico para café no Paraná (in Portuguese). Londrina: IAPAR; 1991. p. 14 (Informe da Pesquisa, 97)

[3] Höfig P, Araujo-Junior CF. Classes de declividade do terreno e potencial para mecanização no estado do Paraná (in Portuguese, with English abstract). Coffee Science. 2015;**10**:195-203. DOI: 10.25186/cs.v10i2.833

[4] Hillel D. Tillage and soil structure management. In: Hillel D, editor. Applications of Soil Physics. New York: Academic; 1980. pp. 234-244

[5] Dias Junior MS, Pierce FJ. Revisão de Literatura. O processo de compactação do solo e sua modelagem (in Portuguese, with English abstract). Revista Brasileira de Ciência do Solo. 1996;**20**:175-182

[6] Castro Filho C. Effect of liming on characteristics of a Brazilian Oxisol at three levels of organic matter as related to erosion [thesis]. Columbus: Ohio State University; 1988

[7] Tavares Filho J, Barbosa GMC, Ribon AA. Water-dispersible clay in soils treated with sewage sludge. Brazilian Journal of Soil Science.2010;**34**(5):1527-1534. DOI: 10.1590/S0100-06832010000500005

[8] Kjaergaard C, Hansen HCB, Koch CB, Villholth KG. Properties of water-dispersible colloids from macropore deposits and bulk horizons of an Agrudalf. Soil Science Society of America Journal. 2004;**68**(5):1844-1852. DOI: 10.2136/sssaj2004.1844

[9] Castro Filho C, Logan TJ. Liming effects on the stability and erodibility of some brazilian oxisols. Soil Science Society of America Journal. 1991;**55**:1407-1413. DOI: 10.2136/sssaj1991.03615995005500050034x

[10] Fontes MPF, Gjorup GB, Alvarenga RC, Nascif PGS. Calcium salts and mechanical stress effects on water-dispersible clay of oxisols. Soil Science Society of America Journal. 1995;**59**(1):224-227. DOI: 10.2136/sssaj1995.03615995005900010034x

[11] Ferreira MM, Fernandes B, Curi N. Influência da mineralogia da fração argila nas propriedades físicas de latossolos da região sudeste do Brasil (in Portuguese, with English abstract). Brazilian Journal of Soil Science. 1999;**23**:515-524. DOI: 10.1590/S0100-06831999000300004

[12] Benites VM, Mendonça ES. Propriedades eletroquímicas de um solo eletropositivo influenciadas pela adição de diferentes fontes de matéria orgânica (in Portuguese, with English abstract). Brazilian Journal of Soil Science. 1998;**22**:215-224. DOI: 10.1590/S0100-06831998000200006

[13] GMdeC B, JFde O, Miyazawa M, Ruiz DB, Tavares Filho J. Aggregation and clay dispersion of an Oxisol treated with swine and poultry manures. Soil & Tillage Research. 2015;**146**:279-285. DOI: 10.1016/j.still.2014.09.022

[14] Lipiec J, Czyz EA, Dexter AR, Siczek A. Effects of soil deformation on clay dispersion in loess soil. Soil & Tillage Research. 2018;**184**:203-206. DOI: 10.1016/j.still.2018.08.005

[15] Guimarães Júnnyor WS, Diserens E, De Maria IC, Araujo-Junior CF, Farhate CVV, de Souza ZM. Prediction of soil stresses and compaction due to agricultural machines in sugarcane

cultivation systems with and without crop rotation. Science of the Total Environment. 2019;**681**:424-434. DOI: 10.1016/j.scitotenv.2019.05.009

[16] Diserens E, Battiato A, Sartori L. Soil compaction, soil shearing and fuel consumption, TASC V3.0—A practical tool for decision-making in farming. In: International Conference of Agricultural Engineering; 6-10 July; Zürich ETH, Switzerland. 2014

[17] Araujo-Junior CF, Dias Junior MS, Guimarães PTG, Alcântara EM. Capacidade de suporte de carga e umidade crítica de um Latossolo induzida por diferentes manejos (in Portuguese, with English abstract). Brazilian Journal of Soil Science. 2011;**35**(1):115-131. DOI: 10.1590/S0100-06832011000100011

[18] Caramori PH, Caviglione JH, Wrege MS, Gonçalves SL, Faria RTF, Androcioli Filho A, et al. Zoneamento de riscos climáticos para a cultura de café (*Coffea arabica* L.) no estado do Paraná (in Portuguese, with English abstract). Brazilian Journal of Agrometeorology. 2001;**9**(3):486-494

[19] Siqueira R, Androcioli Filho A, Pavan MA, Chaves JCD. Densidade de plantio, podas dos primeiros ramos e produção de duas cultivares de café e do híbrido Icatu (in Portuguese, with English abstract). Pesquisa Agropecuária Brasileira. 1983;**18**:763-769

[20] Pavan MA, Bingham FT, Pratt PF. Redistribution of exchangeable calcium, magnesium, and aluminum following lime or gypsum applications to a Brazilian Oxisol. Soil Science Society of America Journal. DOI: 10.2136/sssaj1984.03615995004800010006x

[21] Roth CH, Pavan MA, Chaves JCD, Meyer B, Frede HG. Efeitos das aplicações de calcário e gesso sobre a estabilidade de agregados e infiltrabilidade de água em Latossolo Roxo cultivado com cafeeiros (in Portuguese, with English abstract). Brazilian Journal of Soil Science. 1986;**10**:163-166

[22] Roth CH, Pavan MA. Effects of lime and gypsum on clay dispersion and infiltration in samples of a Brazilian Oxisol. Geoderma. 1991;**48**:351-361. DOI: 10.1016/0016-7061(91)90053-V

[23] Serafim ME, Oliveira GCde, Oliveira ASde, Lima JM, Guimarães PTG, Costa JC. Sistema conservacionista e de manejo intensivo do solo no cultivo de cafeeiros na região do alto São Francisco, MG: um estudo de caso (in Portuguese, with English abstract). BioScience Journal. 2011;**27**:964-977

[24] Serafim ME, de Oliveira GC, Vitorino ACT, Silva BM, Carducci CE. Qualidade física e intervalo hídrico ótimo em Latossolo e Cambissolo, cultivados com cafeeiro, sob manejo conservacionista do solo (in Portuguese, with English abstract). Revista Brasileira de Ciência do Solo. 2013. DOI: 10.1590/S0100-06832013000300020

[25] Carducci CE, de Oliveira GC, Curi N, Heck RJ, Rossoni DF, de Carvalho TS, Costa AL. Gypsum effects on the spatial distribution of coffee roots and the pores system in oxidic Brazilian Latosol. Soil & Tillage Research. 2014. DOI: 10.1016/j.still.2014.09.015

[26] Auler AC, Pires LF, Caires EF. Surface and incorporated liming effects on clay dispersion, water availability, and aeration capacity of a Dystrudept soil. Bragantia. 2017. DOI: 10.1590/1678-4499.290

[27] Chaves JCD, Pavan MA, Miyazawa M. Redução da acidez subsuperficial em coluna de solo (in Portuguese, with English abstract). Pesquisa Agropecuária Brasileira. 1988;**23**(5):469-476

Section 3

Harvest, Post-Harvest and Coffee Quality

Chapter 4

The Harvest and Post-Harvest Management Practices' Impact on Coffee Quality

Mesfin Haile and Won Hee Kang

Abstract

Coffee is one of the most important agricultural commodities in the world. The coffee quality is associated with pre-harvest and post-harvest management activities. Each step starting from selecting the best coffee variety for plantation until the final coffee drink preparation determines the cupping quality. The overall coffee quality influenced by the factors which involve in changes the physicochemical properties and sensorial attributes, including the post-harvest operations. The post-harvest processing activities contribute about 60% of the quality of green coffee beans. The post-harvest operations include pulping, processing, drying, hulling, cleaning, sorting, grading, storage, roasting, grinding, and cupping. This chapter comprises the harvest and post-harvest operations of coffee and their impacts on coffee quality.

Keywords: digestive bioprocessing, coffee cherry, caramelization, Maillard reaction, speciality coffee

1. Introduction

Coffee trees are widely grown in tropical and subtropical regions of Africa, Southeast Asia, and South America [1]. The world annual coffee production estimated 158.6 million 60-kg bags as of 2017/2018, up from 148.6 million 60-kg bags in 2014/2015. South America, Asia and Oceania, Mexico and Central America, and Africa produced as presented, respectively, 81.5, 47.7, 21.7, and 17.8 million 60-kg bags of coffee. The genus *Coffea* belongs to the *Rubiaceae* family and holds more than 90 different species. However, only *Coffea arabica*, *Coffea canephora*, and *Coffea liberica* are commercially important [1]. Arabica coffee accounts for about 64%, while *C. canephora* accounts for about 35% of the world's production; other species with not much commercial value like *Coffea liberica* and *Coffea excelsa* represent only 1% [2]. The quality of coffee is affected by a series of multiple factors. In broad categories, two factors affect coffee quality, namely, pre-harvest and post-harvest factors [3]. The pre-harvest factors set approximately 40% of the sensory attributes and physical and chemical properties of the coffee beans and the remaining 60% of the coffee quality established by the post-harvest processing [4]. Following harvesting, coffee cherries go through a complex series of post-harvest processing steps to be in a more stable, transportable, and roastable form. The initial post-harvest processing steps have a significant role in ensuring the safe changes of the perishable coffee cherries into moderately stable green coffee beans. These green coffee beans have a moisture content of 10–12% to avoid undesired fermentation [5]. The popularity of the coffee product is associated with its distinctive organoleptic properties. Post-harvest management activities conducted to obtain suitably dried coffee beans for roasting and significantly contribute to the quality of the coffee beverage [6]. Post-harvest processing changes the chemical composition of green coffee beans that directly or indirectly influences the quality and end products [3, 7, 8]. These activities include a series of complicated steps including cherry harvesting, de-pulping, fermenting, drying, storage, and others. The number of activities varies according to the type processing method. Right after the on-farm post-harvest process is finished, the coffee can be brought to the coffee industry where the semi-manufactured or complete products are made for commercialization [9].

2. Coffee harvesting and its impact on coffee quality

Production of speciality coffee needs a proper plan for harvesting the coffee cherries as it gives good economic returns for producers. The time of harvest varies in different places. According to the processing method to be implemented, harvesting the coffee cherries without causing damage to the tree is an important task. In most of the coffee-producing countries, coffee is harvested once per year. As the coffee cherries mature, the coffee fruit contains suitable chemical compositions which lead the fruit to the best quality [10]. The coffee fruit also contains volatile compounds that are responsible for the aroma and flavour properties of the coffee. These compounds are present at a very low amount at the early stage of the coffee cherry, but later on, it increases as the coffee transformed to the maturity stage [10]. There are two strategies (strip and selective picking) for harvesting the coffee cherries, which are widely used.

2.1 Strip picking

This strategy is usually done by machinery or by hand. The whole coffee cherries are harvested at one time. The harvested coffee may not achieve the desired quality due to the mixture of underripen or overripen coffee cherries. In order to use machinery for harvesting the coffee cherry, the following factors are critically important such as the topography, inclination, spacing, alignment, and the height of

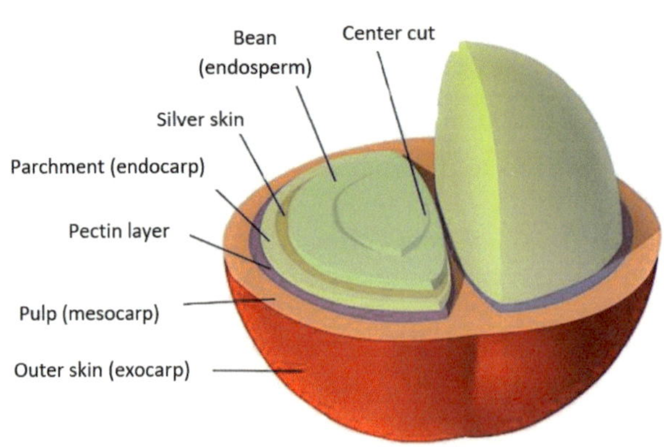

Figure 1.
The coffee cherry anatomy.

Components	%
Ether extract	0.48
Crude extract	21.4
Crude protein	10.1
Ash	1.5
Nitrogen-free extract	31.3
Tannins	7.8
Pectin substance	6.5
Non-reducing sugars	2.0
Reducing sugars	12.4
Chlorogenic acid	2.6
Caffeine	2.3
Total caffeic acid	1.6

Adapted from Gathuo et al. [13].

Table 1.
Composition of coffee pulp.

Components	Concentration (g/L)
Glucose	35.65
Galactose	37.67
Lactose	1.06
Proteins	0.119
Syringaldehyde	0.610
Minerals	**(mg/L)**
Sulphur	30.19
Calcium	37.08
Potassium	239.8
Magnesium	10.05
Phosphorus	41.55
Sodium	7.18
Iron	0.65
Copper	2.45
Zinc	0.14
Manganese	0.07
Boron	0.16
Barium	0.02
Arsenic	0.47
Lithium	0.01
Silicon	1.58
Strontium	0.07

Adapted from Pérez-Sariñana et al. [14].

Table 2.
Chemical composition of coffee mucilage.

Figure 2.
The cherries of different varieties of Coffea arabica.

the plant [11]. Considering these factors, the producer should choose the right harvesting system that suits their crop.

2.2 Selective picking

The major challenges in the coffee sector are obtaining uniform maturity, and at the same time, it is the procedure to provide consistent quality coffee. It is important to note that in most flowering plants the climatic condition during the growing period can change the number of flowering plants which affect the maturation uniformity [12]. In this harvesting system, only the ripened coffee cherries are harvested selectively by hand from the whole tree or branches. The harvested coffee meets the standards due to the uniformity of the cherries. This strategy needs several picking rounds and is labour-intensive. Considering the advantage and disadvantages of both harvesting strategies is the key for coffee growers. The coffee cherry has different layers that surround the beans, such as skin, pulp, mucilage, parchment, silver skin, and finally, the coffee beans (**Figure 1**).

The pulp and mucilage are rich in nutrients, and its chemical compositions are presented in **Tables 1** and 2. These days the coffee pulps are being used for making a coffee wine by fermenting the coffee pulp. So far, there are two patents already registered in Korea. The ripe coffee cherries have colour ranges from bright red to deep red and yellow, depending on the plant variety and for ripe cherries. The different varieties of arabica coffee cherries are presented in **Figure 2**.

3. Post-harvest operations and their impact on coffee quality

3.1 Coffee processing methods

After harvesting of the fruits, green coffee beans are obtained by one of three different methods known as dry, wet, and semi-dry processing [15]. Commonly, there are three different coffee processing methods (**Figure 3**). These methods are wet, dry, and semi-dry processing, and recently, digestive bioprocessing is practised on a small scale to produce the world's most expensive coffee (kopi luwak and black ivory coffee). Although all methods aim at removing the fruit flesh of coffee cherry, they do it in different ways [16]. After harvesting, the coffee cherry follows washing with water to separate floaters (overripe coffee cherries, undeveloped coffee cherries, sticks and leaves). Those processing methods are briefly described below.

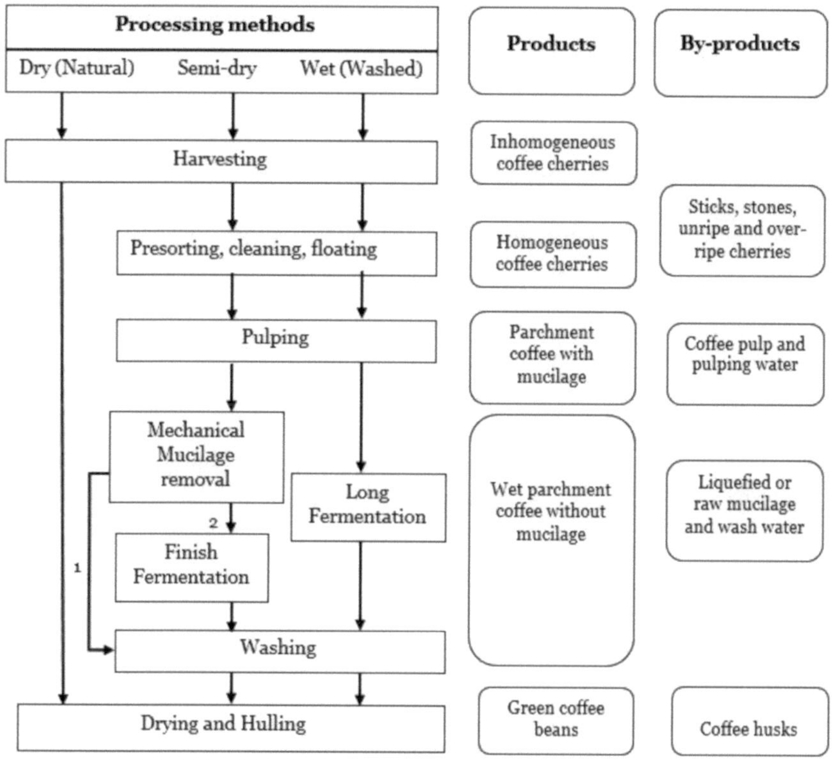

Figure 3.
Coffee processing methods.

3.1.1 Wet processing

This processing method demands the use of some particular facilities and ample amounts of water. When this method adequately implemented, it maintained the inherent quality of coffee beans and produced uniform green coffee beans with minimum defectives. Arabica coffee is widely obtained using wet-processing method except in some countries like Ethiopia, Brazil, and Yemen where they processed in both wet- and dry-processing system, although few percentages of robusta coffee are obtained in this method. Right after harvesting, the coffee cherries are washed and then de-pulped using a machine. The coffee mucilage has pectin substances (polysaccharides), and it has a sticky nature and is challenging to get rid of using water. After de-pulping, chemicals mostly enzymes used to remove mucilage or the natural fermentation proceeds depend upon the environmental condition and fruit ripeness [15]. Natural plant enzymes are present in the coffee fruit that facilitates the degradation of mucilage; however, it is not enough for a full and sufficient process [17]. Microorganisms are responsible for degrading the polysaccharide substances from the parchment coffee. Finally, after fermentation the end product is a "parchment" or "washed" coffee [18]. The type of coffee processing methods (wet and dry) that are performed to obtain the green beans determined the flavour properties and created a typical quality difference [3, 19]. Several reports also stated the final cup quality and chemical composition are also defined by the adopted method in wet processing [20–22]. The effect of different modified wet-processing methods on the final quality of coffee is evaluated and compared by Gonzalez-Rios et al. [21]. Coffee obtained using wet-processing method is considered as a high-quality coffee and received a higher price compared to dry-processed (natural) coffee [23]. A study showed that during processing, several metabolic activities are exhibited in green coffee beans [24, 25]. These metabolic changes occur mainly because of the germination processes, [26] and stress metabolism is a cause for notable differences in the chemical composition of the green coffee beans and thereby determines the quality [27].

As mentioned above, fermentation is a crucial step in wet processing. Diverse microorganisms are existing during fermentation in wet processing [28]. The reason for an excellent microbial population during wet-processing method is because of the nutrient-rich pulp and mucilage. However, microbial heterogeneity subjects to distinction depending upon geographical features, the composition of the coffee fruit, and the fermentation methods [29–31]. These microorganisms are consuming the nutrients in the pulp and mucilage and are responsible for producing different metabolites and organic acids, which are then stored in the coffee beans and may affect the coffee quality. The variations of microorganisms diversity and environment may lead to the difference of the type of these organic acids and metabolites or the amount to which they produced and therefore provide different unique coffee qualities [32, 33].

3.1.2 Semi-dry (semi-wet) processing

The semi-dry (pulped natural) processing is an intermediate process between wet and dry processing. Like wet processing, the coffee exocarp and significant portion of mesocarp are separated while de-pulping is operated. However, unlike wet-processing method, the sticky part (mucilage) remains and is allowed to dry on the parchment instead of complete removal of it with further fermentation until 11–12% moisture content is achieved [34]. After drying the parchment coffee in well-aerated raised beds or cement patios, the coffee bean is then separated from its parchment using machine or any locally available materials like wooden mortar in small scale farms. Despite having less body compared to naturally processed (dry) coffee beans, the cup quality of semi-dry-processed coffee has a bright and clean cup and is somewhat similar to wet-processed beans [35, 36]. A comparison of the primary and secondary metabolites of coffee, which processed by the wet and semi-dry (semi-wet) method was comprehensively investigated [37]. These metabolites are supposed to be very important, because during roasting their degradation results in the development of those nonvolatile or volatile compounds essential for the formation of coffee properties, pigmentation, bitterness, body, astringency, sweetness, aroma, and so on [37–39]. A very similar result between semi-wet- and wet-processed coffees was found regarding the distributions of nine chlorogenic acid classes (3-CQA, 4-CQA, 5-CQA, 3-FQA (feruloylquinic acid), 4-FQA, 5-FQA, 3,4-di-CQA, 3,5-di-CQA, and 4,5-di-CQA), but the total CGA content was statistically lower in semi-wet-processed coffee except for di-CQA [40]. The reason for low CGA content in semi-washed beans might be associated with soaking water for a shorter time which leads to the loss of water-soluble components due to leaching and fermentation [14, 41]. Generally, semi-dry-processing methods are more regularly practised in only Brazil and some parts of Sulawesi and Sumatra.

3.1.3 Dry processing

Dry processing is the oldest, cheapest, and simple processing method. The coffee beans that are obtained using dry-processing method are usually called unwashed (natural) coffee. In this processing method, to get the green coffee beans, the harvested cherry is dried in the sun or other mechanical drier and followed by separating the dried outer parts [22]. Sun drying is a lengthy process, and 95% of

Arabica coffee from Brazil, Ethiopia, Haiti, Indonesia, Paraguay, India, and Ecuador are sun-dried [42]. Few steps are involved in this processing method compared to wet- and semi-dry-processing methods. It is a time-consuming processing as the whole coffee cherry takes some time to dry compared to drying the parchment coffee in wet processing. Dry processing has a high risk of secondary fermentation because of the mucilage, which is very hygroscopic remains with the coffee cherry [43]. The fruits are spread over thin layer and regularly raked to maintain uniform temperature from the top-bottom layer. The drying step in anywhere may take from 10 days to 3 weeks. However, the geographical location and seasons affect the drying process. In some big farms, the mechanical drier is used to speed up the drying process. Because of the existing lowland high temperature and condensation effect, it speeds up the drying process of coffee beans more quickly at the surface (cement or bricks) than raised bed, which is made of mesh wire or bamboo mats. Drying is the most critical step in this processing method because it affects the final quality of the coffee. The overdried fruits have brittle characteristics and produce many defective (broken) during hulling; however, on the other side, under-dried fruits are highly exposed to deteriorations because of fungus and bacteria growth [22]. Tadesse et al. [44] reported that the number of imperfect beans with musty, earthy and greenish colour coffee defects frequently found in dry processed coffee. The bean size and roast volume of dry-processed coffee beans were larger, whereas the wet-processed coffee beans have a higher moisture content [44]. However, concerning the physical parameters (colour, shape, size) dry-processed coffee beans are no longer regarded as better than wet-processed. Coffee beans that are obtained with dry-processing method are heavy in body, smooth, and sweet and have complex characteristics.

3.1.4 Digestive bioprocessing

The world most expensive coffees are obtained through digestive bioprocessing method. This processing method is rare and applied on a small scale. Digestive bioprocessing is a way of passing the coffee cherry through the animal intestine or digestive tract, which is the coffee beans exposed for acids, enzymes, and fermentation treatments. The coffee cherries are eaten by the civet cat and passed through the digestive system to produce civet coffee (kopi luwak) in Indonesia [45]. Similarly, elephant dung coffee (black ivory coffee), produced after the coffee cherries passed through the elephant's digestive tract, which is commonly produced in Thailand. During civet digestion, the breaking down of proteins creates a unique flavour and aroma of civet coffee [45]. The annual production of kopi luwak is less than 500 pounds, and the price is 600 dollars (Canadian) per pound; it leads the indisputable status of being the most expensive and rarest coffee in the world. Currently, we are working on analysing the elephant dung coffee, and it is compared with other coffees that are processed with commonly known processing methods (dry and wet processing) (data not shown). The experiments were conducted in Nepal. The coffee cherries were harvested and fed by the elephants. Then the parchment coffee (the skin and pulp completely digested) and coffee cherry (the pulp is not digested) were collected from the elephant dung (**Figure 4**), and washing, drying, and other processes to obtain the green beans are continued. The black ivory coffee has a price tag of 1000 USD per kilogramme bean. This price makes it one of the world's expensive coffees [46].

3.2 Drying

Drying is the most critical steps in coffee processing methods. Drying is performed to reduce the moisture contents of coffee bean to the required level

Preparation of coffee cherries to feed the elephant

Elephant dung including the coffee cherries and parchment coffee

Parchment coffee (The pulp is completely digested)

Coffee cherries (The pulp is not digested)

Figure 4.
Elephant dung coffee preparation in Nepal.

(10–12%) and to separate the parchment from the coffee beans easily. Sufficient drying is a crucial step to avoid the developments of moulds, which leads to significant losses and affects coffee quality. The drying methods were significantly affected the amount of low molecular weight (LMW) carbohydrates that present in the green coffee beans [47]. Several researchers reported the responses of coffee to various drying processes. The coffee cherry kinetics and drying characteristics under different drying temperature (40, 50, and 60°C) were investigated [47]. The effect of drying and storage conditions on the quality of washed and natural coffee was evaluated by Coradi et al. [48]. This research was focused on the relevance of suitable drying along with sufficient storage conditions to maintain the coffee quality. There are two ways for drying the coffee, such as natural drying using sunlight and mechanical drier.

3.2.1 Sun (natural) drying

This drying method is the most common, widely practised in many countries and the cheapest way of drying the parchment coffee or the whole coffee cherry. This activity is highly dependent on climatic conditions and seasons. The removal of the mucilage and pulps at an early stage in wet-processing reduces the time needed for drying. There are different kinds of sun-drying methods. Coffee is typically dried on large patios made of cement or asphalt concrete with a 0.5–1% slope to drain the water. Using the natural sunlight, drying of coffee takes 7–15 days for parchment coffee and 12–21 days for coffee cherries in patios. This duration varies depending on the weather conditions. The parchment coffee needs special attention than cherries to reduce physical damage (cracking). Raking is required to allow the coffee to dry uniformly, but it should be carried out carefully. In some tropical countries, during the hottest hours of the day, covering sheets used to avoid the cracking of the coffee beans because of overheating. The physicochemical characteristics of coffee that are dried on various types of drying grounds and the thickness of the layers (thin and thick) were evaluated by Reinato et al. [49]. Their result showed the wet-processed parchment coffee dried on thin layer which resulted in the best beverage quality.

3.2.2 Mechanical driers

During the mechanical drying, the beans are heated by hot air that passes through the machine and dries the coffee. Also, it carries away the moistures from the drying coffee. Controlling the drying temperature is the most critical part, and it should not exceed 45°C for whole cherries and 40°C for parchment coffee. Commonly there are two kinds of mechanical drier, such as static and revolving driers. Revolving driers include column driers, vertical driers, rotary driers, cascade driers, and flex driers, and there are tray driers with a stirrer [50]. All these driers can work with fuel oils, gasses, woods, and other energy sources. A report indicated that the best quality coffee was gained when the drying processes were done at two stages, an initial period the drying was performed at low temperature (20°C) and the second stage followed by higher temperature (60°C) [51].

3.3 Hulling and sorting

Hulling is the next step after drying. The dried coffee cherries (dry-processed) or parchment coffee (wet-processed) hulled to remove the covers and get the coffee beans out of it. Once the coffee bean reaches this stage, it means all the essential quality, such as colour and moisture content, has already been achieved. Hulling is done by using different hulling machines, or locally on a small-scale farm, wooden mortar and pestle used to separate the coffee beans from the parchment or dried coffee cherries. Commonly there are two kinds of coffee hulling machines. One is which rubbed off the parchment by friction, and this might create heat. The second type is just cut the parchment and stripped off. It is essential to take care of the coffee beans and avoid physical damage even not to heat the beans during hulling because it affects the colour and taste of the coffee. The last layer that encloses the coffee beans is thin silver skin, and this may be removed or may not be removed during the hulling process. If the coffee beans' silver skin is not separated during hulling; it needs a machine called polishers to separate it and get the green coffee beans. Finally, the green coffee beans are ready to be cleaned and sorted according to colour, size, and density [52].

The hulled coffee beans then undergo sorting processes, which is done by machine and hand. Hand sorting is most widely used, and it requires intensive labour for sorting the coffee beans based on size, colour, and density. The same-size and larger coffee beans get a premium price in the market, and a high percentage of defect may lead to a lower grade, and the price is also low [53]. Sorting is a crucial step because it affects the roasting condition. Uniform size coffee beans should be roasted to achieve uniform roasted beans. Sieve machines are used to screen the coffee beans according to their size. The sieving principle is applied in the sorting machine, which is using a big vibrating flatbed [54]. There are different kinds of machines used to sort coffee beans. Electronic devices are also used to separate the coffee beans by their colours. However, this is not always sufficient to detect and separate the good and bad coffee beans. After sorting is properly done, the coffee beans are then packed with the right packaging materials and transferred to the storage house.

3.4 Storage

Right after the coffee beans are graded, they have to be kept in a storage house until they shipped and sold in the market. The temperature and relative humidity of the storehouse should be controlled to maintain the coffee quality without losing its intrinsic sensorial characteristics and physical and chemical properties and to store for a longer time. Afonso [55] reported that as the storage relative humidity is higher for an extended time, it decreases the compositions of reducing sugars in green coffee beans. Several research reports are available about the effects of storage conditions on the sensorial quality of coffee beans. He also mentioned that the 60% relative humidity with longer storage causes to cellular degradation and leads to oil leaking, which also contributes to the chemical compositions of green coffee beans. When the storage duration is prolonged, the oil becomes more acidic, and it reduced the quality of the product [54].

3.5 Roasting, grinding, and drewing

The most critical factor in the coffee value chain is roasting, where the physicochemical changes lead to the fulfilment of the roasted coffee characteristics [55]. Roasting is considered the essential steps in the formation of the aroma and flavour properties [56]. The essential reactions during coffee roasting and which are responsible for the colour, volatile compounds, and flavour developments are called Maillard reaction and caramelization. Minerals are vital catalysts in the various biochemical reactions responsible for the formation of different aroma and flavour compounds [57]. Amino acids have an important role in the formation of nitrogen/sulphur heterocyclic compounds called melanoidins during roasting because of Maillard reaction or caramelization, which are considered as crucial compounds for flavour development [58]. When the roasting temperatures are higher than 200°C, the precursors in green coffee are transformed into roasted coffee constituents, which lead to the development of diverse aroma test and colour [59]. However, the coffee's intrinsic quality is predetermined in the green bean by its precursor composition, and the roaster only can unlock the full potential by applying the appropriate and optimised roasting conditions. Optimising the appropriate roasting conditions is undoubtedly the most critical ways for achieving the desirable coffee aroma [60]. Fobe and his co-workers [61] reported that as the roasting time is extended, the following changes occurred: the sugar content is reduced and then raised; caffeine contents showed insignificant changes; protein continuously decreased; free fatty acid improved; and unsaponifiable compounds declined. Another report mentioned that the lipid and organic acid increased, while the trigonelline and caffeine content showed almost unchanged [60]. During roasting net losses of matters in the form of water vapour, CO_2, and volatile compounds are exhibited. The roasting duration and the final temperature of the coffee beans determine the development of flavour compounds [62]. A different flavour profile may occur because of the time–temperature and roasting conditions, even though the same coffee beans and roaster are used [63]. The physical properties and kinetics of aroma development of coffee showed differences as the coffee is roasted for a shorter time at high temperature compared to beans roasted for a longer time at low temperature [64, 65]. Fast roasting generates more soluble solids and causes less degradations of CGA, and the loss of volatile was lower [66]. However, the fast roasted coffee is considered affected by lipid oxidation because of high migration of oil from the inner part of the coffee beans to the surface [1]. The determining factors of the final coffee quality are roasting profile, roasting degree, and the technology used for coffee roasting. There are different kinds of roasting conditions. The most common roasting types are discussed below.

3.5.1 Light roast

The lightly roasted coffee developed light brown colour, and it is preferred to make a coffee with a mild body. It has light fragrant, floral or fruity coffee notes. In this roasting condition, the coffee beans should not develop oil on their surface. It is also characterised with pronounced acidity and high caffeine content. During light roasting, the internal temperature of the coffee bean reached approximately 204°C (**Figure 5A**).

3.5.2 Medium roast

Medium roasted coffee developed medium brown colour, and the surface of the coffee beans should not have oils. It is characterised by balanced aroma flavour and acidity. Traditionally this is the most preferred roasting condition. It has medium caffeine content. To achieve this roasting condition, the internal temperature of the coffee beans reached approximately 215°C (**Figure 5B**).

Figure 5.
The most common types of roasting conditions (A, light roast; B, medium roast; C, medium-dark roast; D, dark roast).

3.5.3 Medium-dark roast

Medium-dark roasted coffee has a dark brown colour, and the surface of roasted beans developed oils. It is characterised by fully bodied deep flavoured and little spicy notes. It has low caffeine content. This roasting condition is attained when the internal temperature of the coffee bean reached nearly 229°C (**Figure 5C**). Overall, roasting coffee in medium to dark conditions causes an increase in ketones, esters/lactones, aldehydes, aliphatic acids, and aromatic acids, but a reduction in caffeine content is observed [67–69].

3.5.4 Dark roast

It has nearly a dark colour and produced oils on the surface of the coffee beans. The darker the coffee beans, the less the acidity. It is also characterised by very low caffeine content with heavy mouthfeel, strong flavour, bitter, and burnt or smoky notes. When the coffee is roasted in the dark condition, the internal temperature of the coffee beans reached approximately 246°C (**Figure 5D**).

Grinding is the next step after roasting. In this step, the coffee beans are crushed and changed into powder at different particle size. It is done using a grinding machine (electrical) or using mortar. However, to achieve a uniform ground size, an electrical grinder is the best choice. Blenders can also be used for grinding coffee beans, and the larger particle size and the broadest distributions can be seen while blending the coffee beans [70]. The grinding size is usually measured by using sieve analysis. According to the types of coffee to be prepared, the particle size or grinding particle is the most important. There are many types of grinder with the adjuster to regulate the particle size of coffee grounds. During grinding the important volatile compounds are dispersed into the surroundings from the powdered coffee, so it must be executed right before the intended brewing (1–12 min after grinding) to maintain the volatile compounds [71, 72].

The coffee brewing and extraction methods differ according to the personal preference, the culture, geographical, and social and financial factors. All these brewing and extraction methods of coffee vary depending on the time of extraction, the pressure, the brewing tool, the temperature of the water, and the extraction amounts. The brewing water quality (electrical conductivity) is another important aspect for maintaining the original coffee flavour and test. The size of the coffee ground determines the rate and the total amount of extraction. If the coffee is too fine (for example super-fine) extraction will not happen as water cannot pass through, and if it is too course the grinds will be under-extracted. It depends on the coffee ground size, the coffee brewing methods also vary. Mostly for French press coffee, a coarse coffee ground is preferred, whereas the fine ground coffee is used to make the espresso coffee.

4. Conclusion

The pre-harvest and post-harvest activities starting from selecting the best quality coffee, processing, drying, hulling, storage, roasting, grinding, and brewing can influence the coffee quality. The chemical compositions and physical properties of coffee beans are affected by different factors such as environment, genetics, agronomic activities, harvesting, and post-harvest operations. Except for the genetic and environmental factor, the post-harvest process can be done in a controlled manner to maintain the chemical and physical properties of coffee and thereby maintain its quality. Generally, post-harvest activity determines the quality of green and/or roasted coffee beans. Each step of the post-harvest activities can cause a significant quality loss and lead to a lower market price.

Conflicts of interest

The authors declare no conflict of interest.

Author details

Mesfin Haile[1] and Won Hee Kang[1,2]*

1 Department of Horticulture, Kangwon National University, Chuncheon, Republic of Korea

2 Convergence Program of Coffee Science, Kangwon National University, Chuncheon, Republic of Korea

*Address all correspondence to: whkang@kangwon.ac.kr

References

[1] Schenker SR. Investigations on the Hot Air Roasting of Coffee Beans. Swiss Federal Institute of Technology: Technische Wissenschaften ETH Zürich, Nr.; 2000

[2] Rubayiza AB, Meurens M. Chemical discrimination of arabica and robusta coffees by Fourier transform Raman spectroscopy. Journal of Agricultural and Food Chemistry. 2005;**53**(12): 4654-4659. DOI: 10.1021/jf0478657

[3] Wintgens JN. Factors influencing the quality of green coffee. In: Coffee: Growing, Processing, Sustainable Production. A Guidebook for Growers, Processors, Traders and Researchers. Weinheim, Germany: Wiley-VCH; 2009. pp. 797-817

[4] Musebe RI, Agwanda CH, Mekonen MI. Primary coffee processing in Ethiopia: Patterns, constraints and determinants. African Crop Science Conference Proceedings. 2007;**8**: 1417-1421

[5] Hameed A, Hussain SA, Ijaz MU, Ullah S, Pasha I, Suleria HA. Farm to consumer: Factors affecting the organoleptic characteristics of coffee. II: Postharvest processing factors. Comprehensive Reviews in Food Science and Food Safety. 2018;**17**(5):1184-1237. DOI: 10.1111/1541-4337.12365

[6] Huch M, Franz CM. Coffee: Fermentation and microbiota. In: Advances in Fermented Foods and Beverages. Elsevier, Woodhead Publishing; 2015. pp. 501-513. DOI: 10.1016/B978-1-78242-015-6.00021-9

[7] Bhumiratana N, Adhikari K, Chambers E IV. Evolution of sensory aroma attributes from coffee beans to brewed coffee. LWT- Food Science and Technology. 2011;**44**(10):85-92

[8] Sunarharum WB, Williams DJ, Smyth HE. Complexity of coffee flavor: A compositional and sensory perspective. Food Research International. 2014;**62**:315-325. DOI: 10.1016/j.foodres.2014.02.030

[9] Pereira G, Soccol VT, Brar SK, Neto E, Soccol CR. Microbial ecology and starter culture technology in coffee processing. Critical Reviews in Food Science and Nutrition. 2017;**57**(13): 2775-2788. DOI: 10.1080/10408398.2015.1067759

[10] Pimenta CJ, Pereira MC, Chalfoun SM, Angélico CL, de Carvalho GL, Martins RT. Chemical composition and quality of coffee (*Coffea arabica* L.) collected at different harvesting times. Revista Brasileira de Armazenamento. 2008;Especial **10**:29-35

[11] Ventureli I, Patto GJ, Beli E, Mandelli I. Comparação de custos entre colheita mecanizada e semi-mecanizada de café: Um estudo de caso em uma propriedade no município de Andradas–MG. Revista ADM. 2016;**16**(20):93-104

[12] Pimenta CJ, Angélico CL, Chalfoun SM. Challengs in coffee quality: Cultural, chemical and microbiological aspects. Ciência e Agrotecnologia. 2018;**42**(4):337-349. DOI: 10.1590/1413-70542018424000118

[13] Gathuo B, Rantala P, Määttä R. Coffee industry wastes. Water Science and Technology. 1991;**24**(1):53-60. DOI: 10.2166/wst.1991.0009

[14] Pérez-Sariñana BY, De León-Rodriguez A, Saldaña-Trinidad S, Joseph SP. Optimization of bioethanol production from coffee mucilage. BioResources. 2015;**10**(3):4326-4338

[15] Teixeira AA, Brando CHJ, Thomaziello RA, Teixiera R. In: Illy A, Viani R, editors. Espresso Coffee: The Science of Quality. Italy: Elsevier Academic Press; 1995. p. 91-95; 197-198

[16] Trugo LC, Macrae RA. Chlorogenic acid composition of instant coffee. The Analyst. 1984;**109**:263-266. DOI: 10.1039/AN9840900263

[17] Agate AD, Bhat JV. Role of pectinolytic yeasts in the degradation of mucilage layer of *Coffea robusta* cherries. Applied and Environmental Microbiology. 1966;**14**(2):256-260

[18] Smith AR. In: Clifford MN, Willson KC, editors. A History of Coffee in Coffee: Botany, Biochemistry and Production of Beans and Beverage. Westport, Connecticut, USA: AVI Publishing Company; 1985

[19] Sivetz M, Desrosier NW. Hulling, Classifying, Storing, and Grading Green Coffee Beans. Coffee Technology. Westport, CT: The Avi Publishing Company; 1979. pp. 117-169

[20] Gonzalez-Rios O, Suarez-Quiroz ML, Boulanger R, Barel M, Guyot B, Guiraud JP, et al. Impact of "ecological" post-harvest processing on the volatile fraction of coffee beans: I. Green coffee. Journal of Food Composition and Analysis. 2007;**20**(3–4):289-296. DOI: 10.1016/j.jfca.2006.07.009

[21] Gonzalez-Rios O, Suarez-Quiroz ML, Boulanger R, Barel M, Guyot B, Guiraud JP, et al. Impact of "ecological" post-harvest processing on coffee aroma: II. Roasted coffee. Journal of Food Composition and Analysis. 2007;**20**(3–4):297-307. DOI: 10.1016/j.jfca.2006.12.004

[22] Taveira JD, Borem FM, Da Rosa SD, Oliveira PD, Giomo GS, Isquierdo EP, et al. Post-harvest effects on beverage quality and physiological performance of coffee beans. African Journal of Agricultural Research. 2015;**10**(12):1457-1466. DOI: 10.5897/AJAR2014.9263

[23] Rothfos NB. Coffee Production. Hamburg: Gordian–Max–Rieck; 1985

[24] Selmar D, Bytof G, Knopp SE. New aspects of coffee processing: The relation between seed germination and coffee quality. In: 19th International Coffee Science Conference, ASIC, Trieste (CD-ROM); 2001

[25] Selmar D, Bytof G, Knopp SE, Breitenstein B. Germination of coffee seeds and its significance for coffee quality. Plant Biology. 2006;**8**(02):260-264. DOI: 10.1055/s-2006-923845

[26] Bytof G, Knopp SE, Kramer D, Breitenstein B, Bergervoet JH, Groot SP, et al. Transient occurrence of seed germination processes during coffee post-harvest treatment. Annals of Botany. 2007;**100**(1):61-66. DOI: 10.1093/aob/mcm068

[27] Bytof G, Knopp SE, Schieberle P, Teutsch I, Selmar D. Influence of processing on the generation of γ-aminobutyric acid in green coffee beans. European Food Research and Technology. 2005;**220**(3–4):245-250. DOI: 10.1007/s00217-004-1033-z

[28] Haile M, Kang WH. The role of microbes in coffee fermentation and their impact on coffee quality. Journal of Food Quality. 2019;**2019**:4836709. DOI: 10.1155/2019/4836709

[29] Feng X, Dong H, Yang P, Yang R, Lu J, Lv J, et al. Culture-dependent and-independent methods to investigate the predominant microorganisms associated with wet processed coffee. Current Microbiology. 2016;**73**(2):190-195. DOI: 10.1007/s00284-016-1047-3

[30] Silva SD, Queiroz DM, Pinto FD, Santos NT. Characterization and delimitation of the terroir coffee in plantations in the municipal district of Araponga, Minas Gerais, Brazil. Revista Ciência Agronômica. 2014;**45**(1):18-26. DOI: 10.1590/S1806-66902014000100003

[31] Silva C. Microbial activity during coffee fermentation. In: Cocoa and

Coffee Fermentations. Boca Raton, Florida, USA: CRC Press; 2014. pp. 368-423

[32] Massawe GA, Lifa SJ. Yeasts and lactic acid bacteria coffee fermentation starter cultures. International Journal of Postharvest Technology and Innovation. 2010;**2**(1):41-82. DOI: 10.1504/IJPTI.2010.038187

[33] Silva CF, Vilela DM, de Souza Cordeiro C, Duarte WF, Dias DR, Schwan RF. Evaluation of a potential starter culture for enhance quality of coffee fermentation. World Journal of Microbiology and Biotechnology. 2013; **29**(2):235-247. DOI: 10.1007/s11274-012-1175-2

[34] Ribeiro LS, Miguel MG, Evangelista SR, Martins PM, van Mullem J, Belizario MH, et al. Behavior of yeast inoculated during semi-dry coffee fermentation and the effect on chemical and sensorial properties of the final beverage. Food Research International. 2017;**92**: 26-32. DOI: 10.1016/j.foodres.2016.12.011

[35] Espresso and Coffee Guide: Semi-Washed Processing. 2008. Retrieved from: https://www.espressocoffeeguide.cm/all-about-coffee-2/coffee-processing/semi-washed-processing

[36] Stumptown: Semi-Washed Coffee Processing. Stumptown Coffeeroaster Blog. 2011. Retrieved from: https://www.stumptowncoffee.com/processing/semi-washed/

[37] Farah A. Distribuição nos grãos, importância na qualidade da bebidae biodisponibilidade dośacidos clorogénicos do cafe [PhD thesis] (in Portuguese). Brazil: UFRJ; 2005

[38] Farah A, Monteiro MC, Calado V, Franca AS, Trugo LC. Correlation between cup quality and chemical attributes of Brazilian coffee. Food Chemistry. 2006;**98**(2):373-380. DOI: 10.1016/j.foodchem.2005.07.032

[39] Toci AT, Farah A. Volatile compounds as potential defective coffee beans' markers. Food Chemistry. 2008; **108**(3):1133-1141. DOI: 10.1016/j.foodchem.2007.11.064

[40] Duarte GS, Pereira AA, Farah A. Chlorogenic acids and other relevant compounds in Brazilian coffees processed by semi-dry and wet post-harvesting methods. Food Chemistry. 2010;**118**(3):851-855. DOI: 10.1016/j.foodchem.2009.05.042

[41] Wootton AE. The dry matter loss from parchment coffee during field processing. In: COLLOQUE Scientifique International Sur le Café, 5; Lisbonne, Portugal; 1971. pp. 14-19

[42] Silva CF, Batista LR, Abreu LM, Dias ES, Schwan RF. Succession of bacterial and fungal communities during natural coffee (*Coffea arabica*) fermentation. Food Microbiology. 2008; **25**(8):951-957. DOI: 10.1016/j.fm.2008.07.003

[43] Coste R. Coffee, the Plant and the Product. London: Macmillan Press Limited; 2003

[44] Tadesse FT, Jemal Y, Abebe H. Effect of green coffee processing methods and roasting temperatures on physical and cup quality of Sidama coffee, Southern Ethiopia. Journal of Nutritional Ecology and Food Research. 2016;**3**(1):44-50. DOI: 10.1166/jnef.2016.1119

[45] Marcone MF. Composition and properties of Indonesian palm civet coffee (Kopi Luwak) and Ethiopian civet coffee. Food Research International. 2004;**37**(9):901-912. DOI: 10.1016/j.foodres.2004.05.008

[46] Kleinwächter M, Selmar D. Influence of drying on the content of

sugars in wet processed green Arabica coffees. Food Chemistry. 2010;**119**(2): 500-504. DOI: 10.1016/j.foodchem. 2009.06.048

[47] Corrêa PC, Resende O, Ribeiro DM. Drying characteristics and kinetics of coffee berry. Revista Brasileira de Produtos Agroindustriais. Brazil: SBICafé Biblioteca do Café, Viãosa; 2006;**8**(1):1-10

[48] Coradi PC, Borém FM, Saath R, Marques ER. Effect of drying and storage conditions on the quality of natural and washed coffee. Coffee Science. 2007;**2**(1):38-47. DOI: 10.25186/cs.v2i1.37

[49] Reinato CH, Borem FM, Cirillo MÂ, Oliveira EC. Quality of the coffee dryed on grounds with different surfaces and thickness layers. Coffee Science. 2012;**7**(3):223-237. DOI: 10.25186/cs. v7i3.327

[50] Ghosh P, Venkatachalapathy N. Processing and drying of coffee—A review. International Journal of Engineering Research and Technology. 2014;**3**(12):784-794

[51] Rolz C, Menchu JF, Arimany E. The fluidized bed drying of coffee: 4e Colloque International sur la chimie des cafés. Amsterdam, ASIC, Paris; 1969. pp. 166-173

[52] Wanyonyi JJ. An Outline of Coffee Processing. Crop Production Division, Nairobi, Kenya: Ministry of Agriculture; 1972

[53] Mutua J. Post Harvest Handling and Processing of Green Coffee in African Countries. FAO Corporate Document Repository. 2000. Available from: http://www.fao.org/DOCREP/003/x6939e03.html

[54] NTCDB. Organic Coffee Production and Processing: Guide Book. Kathmandu, Nepal: National Tea and Coffee Development Board; 2009

[55] Afonso PC. Aspectos Físicos, Fisiológicos e de Qualidade do Café em Função da Secagem e do Armazenamento. Minas Gerais, 2001

[56] Arya M, Rao LJ. An impression of coffee carbohydrates. Critical Reviews in Food Science and Nutrition. 2007; **47**(1):51-67. DOI: 10.1080/10408390600550315

[57] Oestreich JS. Chemistry of coffee. In: Lew M, Hung-Wen L editors. Comprehensive Natural Products. Oxford: Elsevier; Vol. 2, Issue 3; 2010. pp. 1085-1117

[58] Shibamoto T. Heterocyclic compounds in browning and browning/nitrite model systems: Occurrence, formation mechanisms, flavor characteristics and mutagenic activity. In: Charalambous G, Inglett G, editors. Instrumental Analysis of Food. Vol. 1. New York: Academic Press; 1983. pp. 229-278

[59] Dalla Rosa M, Lerici CR, Piva M, Fini P. Processi di trasformazione del caffe: Aspetti chimici, fisici e tecnologici. Nota 5: Evoluzione di alcuni caratteri fisici del caffenel corso di trattamenti termici condotti a temperatura costante. Industrie delle Bevande. 1980;**9**(12):466-472

[60] Buffo RA, Cardelli-Freire C. Coffee flavour: An overview. Flavour and Fragrance Journal. 2004;**19**(2):99-104. DOI: 10.1002/ffj.1325

[61] Fobe LA, Mery JP, Tango JS. Influence of roasting degree on the chemical composition of coffee. In: COLLOQUE Scientifique International Sur le Café, 3; Trieste, Italia; 1967. pp. 2-9

[62] Huschke R. Industrial Coffee Refinement: Process and Systems Engineering from the Green Coffee Reception to Packaging. Munich: SV

Corporate Media, Verlag Moderne Industrie; 2007

[63] Schenker S, Heinemann C, Huber M, Pompizzi R, Perren R, Escher R. Impact of roasting conditions on the formation of aroma compounds in coffee beans. Journal of Food Science. 2002;**67**(1):60-66. DOI: 10.1111/j.1365-2621.2002.tb11359.x

[64] Baggenstoss J, Poisson L, Kaegi R, Perren R, Escher F. Coffee roasting and aroma formation: Application of different time–temperature conditions. Journal of Agricultural and Food Chemistry. 2008;**56**(14):5836-5846. DOI: 10.1021/jf800327j

[65] Gloess AN, Vietri A, Wieland F, Smrke S, Schönbächler B, López JA, et al. Evidence of different flavour formation dynamics by roasting coffee from different origins: On-line analysis with PTR-ToF-MS. International Journal of Mass Spectrometry. 2014;**365**: 324-337. DOI: 10.1016/j.ijms.2014.02.010

[66] Nagaraju VD, Murthy CT, Ramalakshmi K, Rao PS. Studies on roasting of coffee beans in a spouted bed. Journal of Food Engineering. 1997; **31**(2):263-270. DOI: 10.1016/S0260-8774(96)00026-X

[67] Lyman DJ, Benck R, Dell S, Merle S, Murray-Wijelath J. FTIR-ATR analysis of brewed coffee: Effect of roasting conditions. Journal of Agricultural and Food Chemistry. 2003;**51**(11):3268-3272. DOI: 10.1021/jf0209793

[68] Movasaghi Z, Rehman S, ur Rehman DI. Fourier transform infrared (FTIR) spectroscopy of biological tissues. Applied Spectroscopy Reviews. 2008;**43**(2):134-179. DOI: 10.1080/05704920701829043

[69] Wang J, Jun S, Bittenbender HC, Gautz L, Li QX. Fourier transform infrared spectroscopy for Kona coffee authentication. Journal of Food Science. 2009;**74**(5):C385-C391. DOI: 10.1111/j.1750-3841.2009.01173.x

[70] Baggenstoss J, Thomann D, Perren R, Escher F. Aroma recovery from roasted coffee by wet grinding. Journal of Food Science. 2010;**75**(9): C697-C702. DOI: 10.1111/j.1750-3841.2010.01822.x

[71] Akiyama M, Murakami K, Ohtani N, Iwatsuki K, Sotoyama K, Wada A, et al. Analysis of volatile compounds released during the grinding of roasted coffee beans using solid-phase micro extraction. Journal of Agricultural and Food Chemistry. 2003;**51**(7):1961-1969. DOI: 10.1021/jf020724p

[72] Akiyama M, Murakami K, Ikeda M, Iwatsuki K, Kokubo S, Wada A, et al. Characterization of flavor compounds during grinding of roasted rubusta coffee beans. Food Science and Technology Research. 2005;**11**(3): 298-307. DOI: 10.3136/fstr.11.298

Section 4

Chemistry of Coffee

Chapter 5

A Detail Chemistry of Coffee and Its Analysis

Hemraj Sharma

Abstract

This review article highlights the detailed chemistry of coffee including its components; chemical constituents like carbohydrates, proteins, lipids, and caffeine; aromatic principles; oil and waxes; and minerals and acids. The high extent of caffeine can be found in the coffee plants; hence, in the second part of the study, various analytical methods are designed for the proper identification, separation, optimization, purification, and determination of caffeine present in coffee, tea, and marketed coffee. These analytical methods are appropriated for the separation and quantification of caffeine. The various analytical methods include spectroscopy methods like UV, IR, and NMR spectroscopy; chromatographic methods like paper, TLC, column, HPLC, and gas chromatography; and hyphenated techniques like LC–MS, GC–MS, and GC–MS/MS. This article compares and contrasts the amount of caffeine by various analytical methods.

Keywords: caffeine, spectrophotometer, chromatography, hyphenated techniques, electrochemical methods

1. Introduction

Coffee consists of ripe seeds of *Coffea arabica* Linn., belonging to family Rubiaceae. Coffee extracted from coffee bean is also present in crimson fruits is completely removed, and the spermoderm is removed, occasionally. The seeds of botanical genus *Coffea* may be raw, roasted, whole, or ground. The prepared drink through such coffee seeds is also called as coffee. Among 70 species of coffee, only three are cultivated. 75% of the world's production of coffee is provided by *Coffea arabica*, about 25% by *Coffea canephora*, and less than 1% by *Coffea liberica* and others. Generally, coffee is cultivated at the altitude of 1000–2000 [1]. It is indigenous to Ethiopia, Brazil, India, Vietnam, Mexico, Nepal Guatemala, Indonesia, and Sri Lanka.

2. Chemical constituents

The main constituents of coffee are caffeine, tannin, fixed oil, carbohydrates, and proteins. It contains 2–3% caffeine, 3–5% tannins, 13% proteins, and 10–15% fixed oils. In the seeds, caffeine is present as a salt of chlorogenic acid (CGA). Also it contains oil and wax [2].

The following sections will be discussed in detail after acceptance of this short proposal:

IntechOpen

- This article will deal on the types of carbohydrate, protein, lipids, and other chemical constituents in detail.

- This article will review on various analytical methods for the estimation of constituents present in coffee.

Coffee is often used as antioxidants, but more importantly coffee is a good source of chromium and magnesium that assist in controlling blood sugar by ensuring proper usage of insulin.

The main chemical ingredients in coffee beans are given below:

- Caffeine
- Tannin
- Thiamin
- Xanthine
- Spermidine
- Guaiacol
- Citric acid
- Chlorogenic acid
- Acetaldehyde
- Spermine
- Putrescine
- Scopoletin

The carbohydrate content of green and roasted coffee (Santos) was identified and measured. Green coffee contained about 6–7% of sucrose as soluble sugars and low amount of glucose. The soluble sugars of roasted coffee were sucrose, fructose, and glucose. The experiment was also carried out for the isolation of holocellulose fractions of green and roasted coffee.

The holocellulose of green coffee was hydrolyzed by a novel method consisting of anhydrous sulfuric acid and 10% potassium insoluble hydroxide, which was partially solubilized on roasting and results in the following ratio of sugars:

1 L-arabinose/2D-galactose/2D-glucose/6D-mannose. Out of these sugars, the arabinose was easily acid-hydrolyzed. Other coffee constituent analyzed and determined were caffeine, trigonelline, caffeic acid, chlorogenic acid, isochlorogenic acid, and the 10 amino acids. The free amino acids disappeared in roasting. An analytical method was developed for evaluating caffeine on chromatograms [3].

In coffee pulp, condensed tannins are the major phenolic compounds, while in the seeds, phenolic compounds exist primarily as a family of esters formed between hydroxycinnamic acids and quinic acid, collectively recognized as chlorogenic acids (CGA). Green coffee seeds contain up to 14% CGA, which are present in high concentrations and have a greater influence for determining the quality of coffee and play a vital role in the formation of the coffee flavor. The various constituents along with components of coffee are shown in **Table 1**.

Constituent	Components
Soluble carbohydrates	Monosaccharides Fructose, glucose, galactose, arabinose (traces)
Oligosaccharides	Sucrose, raffinose, stachyose
Polysaccharides	Polymers of galactose, mannose, arabinose, glucose Insoluble polysaccharides
Hemicelluloses	Polymers of galactose, arabinose, mannose Cellulose Acids and phenols Volatile acids
Nonvolatile aliphatic acids	Citric acid, malic acid, quinic acid
Chlorogenic acids	Mono-, dicaffeoyl- and feruloylquinic acid Lignin Lipids Wax
Oil	Main fatty acids: N Compounds
Free amino acids	Main amino acids: Glu, Asp, Asp-NH2 Proteins
Caffeine	Traces of theobromine and theophylline Trigonelline Minerals

Table 1.
Constituents along with components of coffee.

3. Carbohydrates

Most of the carbohydrates present, such as cellulose and polysaccharides consisting of mannose, galactose, and arabinose, are insoluble.

4. Lipids

The lipid fraction appears to be very stable, and its composition is given below.
Linoleic acid is the predominant fatty acid, followed by palmitic acid.
Lipid composition.
Triacylglycerols.
Diterpene esters.
Diterpenes.
Triterpene esters.
Triterpenes (sterols).
Unidentified compounds.

5. Acids

The volatile acids include formic acids and acetic acids, while nonvolatile acids include lactic, tartaric, pyruvic, and citric acid. Minor constituents include higher fatty acids and malonic, succinic, glutaric, and malic acids. The degradation products of citric acid are itaconic (I), citraconic (II), and mesaconic acids (III), while fumaric and maleic acids are degraded products of malic acid:

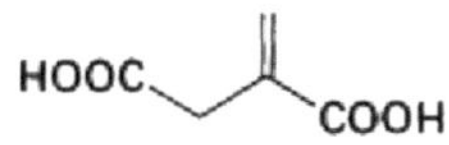

I

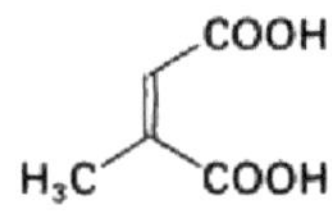

II

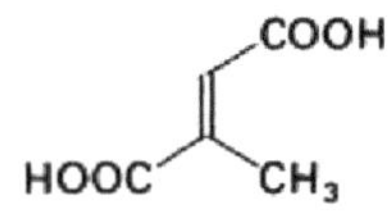

III

Chlorogenic acids are the mainly rich acids of coffee.

6. Trigonelline and nicotinic acid

Green coffee contains trigonelline (N-methylnicotinic acid) up to 0.6% and is 50% decomposed during roasting. The degradants include nicotinic acid, pyridine, 3-methyl pyridine, nicotinic acid, methyl ester, and other compounds.

7. Aromatic principle

The aroma profile of coffee is composed of the following notes: sweet/caramel-like, earthy, sulfurous/roasty, and smoky/phenolic.

8. Minerals

Potassium is major in coffee ash (1.1%), calcium (0.2%), and magnesium (0.2%). The major anions includes phosphate (0.2%) and sulfate (0.1%), along with traces of other elements [4].

9. Caffeine

The best known N compound is caffeine (1,3,7-trimethylxanthine) because of its physiological effects (stimulation of the central nervous system, increased blood circulation, and respiration). It is mildly bitter in taste. 10% of the caffeine and about 6% of the chlorogenic acid are present in a coffee drink. During roasting, the caffeine level in beans is decreased. Synthetic caffeine and caffeine obtained by the decaffeination process are used by the pharmaceutical and soft drink industries. By methylation of xanthine, synthetic caffeine is obtained which is obtained from uric acid and formamide. Medicinally, caffeine is used as a CNS stimulant, usually combined with another therapeutic agent and in analgesic preparations.

Theobromine acts as diuretic and smooth muscle relaxant, but not routinely used. Theophylline is used as smooth muscle relaxant and is frequently dispensed in sustainable formulations to lower the side effects. It is also available as aminophylline (a more soluble preparation containing theophylline with ethylenediamine) and choline theophyllinate (theophylline and choline). The alkaloids may be isolated from natural sources or obtained by total or partial synthesis [5].

The purine alkaloids include caffeine, theobromine, and theophylline as shown in **Figure 1**. They have a limited distribution as alkaloids, but the origins are very close with those of the purine bases like adenine and guanine, fundamental components of nucleosides, nucleotides, and the nucleic acids. Caffeine is mainly consumed in the form of beverages like tea, coffee, and cola and is most widely consumed and socially accepted natural stimulants. Theophylline is much more important as a drug compound because of its muscle relaxant properties, utilized in the relief of bronchial asthma when compared to caffeine, medicinally. The major constituent of cocoa and related chocolate products is theobromine.

Out of four nitrogen atoms, two are supplied by glutamine and a third by aspartic acid. The synthesis of the nucleotides AMP and GMP is by way of IMP and XMP, and the purine alkaloids then branch away via XMP. The loss of phosphate via methylation generates the nucleoside 7-methylxanthosine, which is then released from the sugar moiety. Furthermore, successive methylation on the nitrogen gives caffeine through theobromine, while a different methylation sequence can result in the formation of theophylline (**Table 2**) [6].

AMP = adenosine-5′-monophosphate.
GMP = guanosine-5′-monophosphate.
IMP = inosine-5′-monophosphate.
XMP = xanthosine-5′-monophosphate.

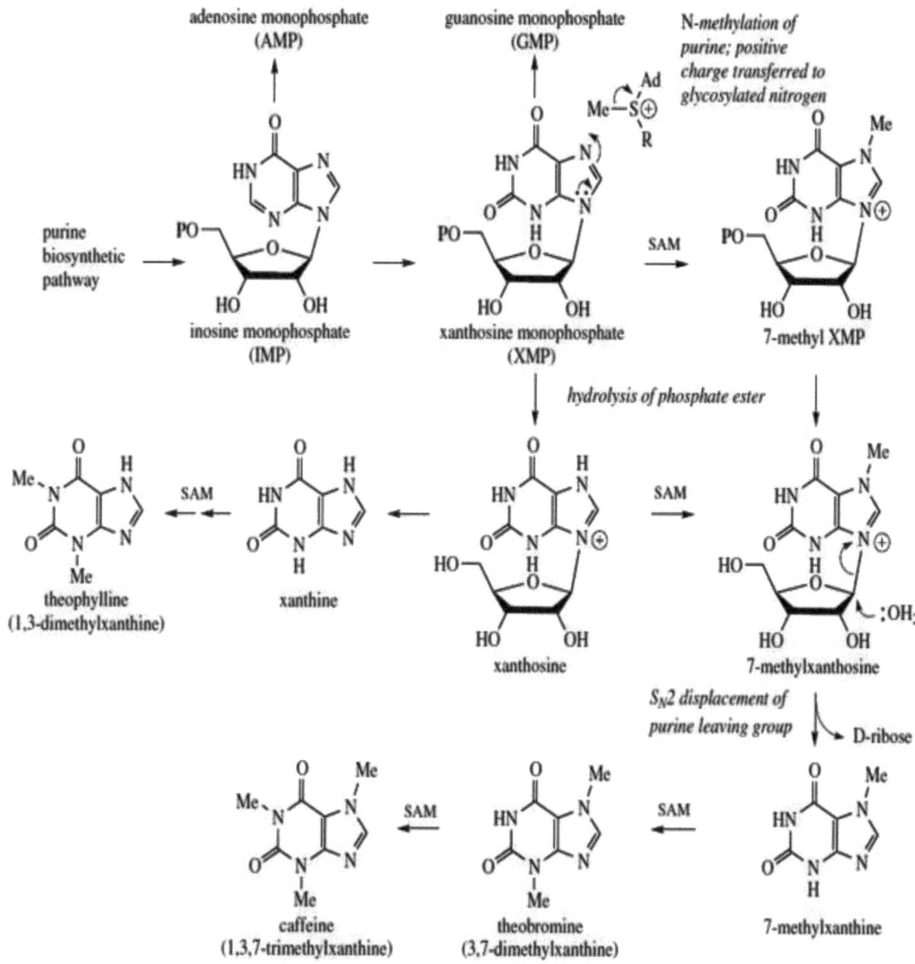

Figure 1.
Chemistry of the purine derivatives.

S.N.	Method	Experiment	Detection	Linearity range	Application	Scientific outcome	Ref.no.
1	UV spectroscopy	Caffeine separated from coffee using paper and TLC and was estimated using spectroscopy	Detection was done at 272 nm	NA	Caffeine from coffee	Good separation	[7]
2.	UV spectroscopy	Caffeine separated from coffee using TLC and was estimated using spectroscopy	Absorbance measured at 274 nm	2–120 μg/ml	Caffeine from tea powder	Good separation	[8]
3.	UV spectroscopy	Method A: simultaneous equation method Method B: isosbestic point method	For method A: absorbance measured at 273 nm For method B: absorbance measured at 259.5 nm	2–32 μg/ml	Tablet containing caffeine and paracetamol	Determination of caffeine in mixture of tablets	[9]
4.	UV spectroscopy	Dual wavelength method	Two wavelengths of 249 and 234 nm were selected for analysis LOD = 0.286 LOQ = 0.863	3–18 μg/ml	Tablet containing caffeine and paracetamol	A new method of determination of caffeine	[10]
5.	HPLC	RP-HPLC comprising C18 column and 24% methanol as mobile phase	UV detector at 272 nm	1–40 ppm	Unroasted coffee and roasted coffee	Unroasted coffee contained 0.89–2.10 (8 samples) Roasted coffee contained 1.03–4.21 (11 samples)	[11]
6.	HPTLC-UV	Silica gel 60F254 as stationary phase and ethyl acetate/methanol (27:3) as mobile phase	UV densitometric remission at 274 nm LOD = 40 ng/zone LOQ = 120 ng/zone	2–14 μg/zone	Caffeine in marketed tea granules	Caffeine in tea samples was found to be 2.145%	[12]
7	HPLC	Zorbax eclipse XDB comprising C8 column as stationary phase and water-tetrahydrofuran-acetonitrile as mobile phase	UV detector at 273 nm LOD = 0.07 LOQ = 0.20	0.2–100 mg/l	Caffeine, theobromine, and theophylline in food, drinks, and herbal products	The recoveries range from 92.00 to 96.8%	[13]

S.N.	Method	Experiment	Detection	Linearity range	Application	Scientific outcome	Ref.no.
8	HPLC and biosensor method	For HPLC: Shimadzu LC10A fitted with a C18 column as stationary phase and acetonitrile and water (10:90%) as mobile phase set at a flow rate of 1 ml min^{-1}	UV detector set at 273 nm	0.01–0.1%w/v	Commercial coffee samples and cola drinks	0.033–0.072%w/v	[14]
		For biosensor: amperometricbiosensor comprising the biological sensing element, transducer, amplification, and detector systems		0.01–0.1%w/v		0.030–0.076%w/v	
9	HPLC	HPLC with solid phase extraction (SPE) HPLC model: Waters 515, with UV detector (REX, Model pHS-25), Visi TM-1 SPE single-sample processor (Supelco) 50 mM KH2PO4 (pH = 2) Acetonitrile and methanol (40:8:2) was used as solvent as well as mobile phase	Caffeine was extracted from green tea, black tea, and coffee and then characterized by melting point, λ max (UV/vis), IR absorption bands, Rf (TLC), and RT (HPLC) Crude caffeine was purified by solid phase extraction	10–60 ppm	Caffeine in tea, coffee, and soft drinks	Crude black tea, green tea, and coffee contained 7.04%, 4.88%, and 13.7% caffeine, respectively, whereas after purification black tea, green tea, and coffee contained 3.34%, 2.24%, and 5.20% pure caffeine	[15]
10.	HPLC and UV	UV/vis spectrophotometer The molar decadic absorption (MDA) coefficients and transitional dipole moment of pure caffeine in water and dichloromethane (DCM) were obtained at 272 and 274.7 nm	MDA was found to be 1115 and 1010 $m^2 mol^{-1}$, respectively, in water and DCM Transitional dipole moments of caffeine in water and in dichloromethane are 10.40×10^{-30} and 10.80×10^{-30} C m, respectively	0.90–1.10% for five samples by HPLC	Caffeine in coffee beans	UV/vis spectrophotometer: five independent measurements were $1.1 \pm 0.01\%$ for Bench Maji, $1.01 \pm 0.04\%$ for Gediyo Yirga Chefe, $1.07 \pm 0.02\%$ for Tepi, and $1.19 \pm 0.02\%$ for Godere, respectively HPLC: measurements were 1.10% for Bench Maji, 1.10% for Gediyo Yirga Chefe, 1.00% for Gomma Limu, and 0.90% for Besema	[16]

S.N.	Method	Experiment	Detection	Linearity range	Application	Scientific outcome	Ref.no.
11	HPLC with DAD	Stationary phase: RP-HPLC (Spherisorb ODS2 column) Mobile phase: 0.01 M phosphate buffer of pH 4	DAD detector at 265 nm LOD = 0.05 μg/ml	0.05–500 μg/ml	Thermal degradation of caffeine in coffee of Brazil and Ivory Coast	For Brazil: green coffee (g/kg of caffeine), 12.36 ± 0.10; roasted coffee, 16.12 ± 0.05 For Ivory Coast: green coffee (g/kg of caffeine), 20.83 ± 0.22; roasted coffee, 25.55 ± 0.185	[17]
12	HPLC	Stationary phase: RP-HPLC C18 Mobile phase: acetonitrile/water (8:92%)	Detection at wavelength of 245 nm.	Varies with each sample	Caffeine and theobromine in coffee, tea, and instant hot cocoa mixes	Instant tea: 32.4–35.0 mg/cup of caffeine Tea bag: 30.2–67.4 mg/cup, 1.0–7.8 mg/cup of caffeine Instant hot cocoa:46.7–67.6 mg/cup of caffeine Ground coffee: 93.0–163.5 mg/cup of caffeine	[18]
13	LC–MS	For LC stationary phase: Spherisorb S5ODS2, 5 μm Mobile phase: formic acid/methanol For MS: ESI source with +ve mode	LOD = 11.9 ng/ml LOQ = 39.6 ng/ml	0.05–25.00 μg/mL	Caffeine, trigonelline, nicotinic acid, and sucrose in coffee	Caffeine values ranged from 843.3 to 930.9 mg/100 g coffee in green and roasted Arabica coffee samples	[19]
14	Electrochemical method	Voltammetric method with CH1760D electrochemical working standard Working electrode: lignin modified glassy carbon electrode Auxiliary electrode: platinum coil Reference electrode: Ag/Agcl	LOD = 8.37×10^{-7} LOQ = 2.79×10^{-6}	$6–100 \times 10^{-6}$ mol/L	Caffeine content in Ethiopian coffee samples	10.78, 8.78, 6.35, 5.85 mg/g caffeine in coffee	[20]

S.N.	Method	Experiment	Detection	Linearity range	Application	Scientific outcome	Ref.no.
15	Electrochemical method	Voltammetric method Working electrode: pencil type graphite carbon electrode Auxiliary electrode: platinum coil Reference electrode: Ag/Agcl electrode	LOD = 9.2 mg/L	0–500 mg/L	Caffeine levels in several tea samples	Caffeine levels in several tea samples yield relative error of 1% in the concentrations	[21]
16	LC–MS/MS	For LC, stationary phase: RP-HPLC C18 Mobile phase: isocratic mobile phase consisting of 0.2% formic acid in distilled water and methanol (80:20, v/v) For MS: spectrometer equipped with an electrospray Ionization mode used to generate positive [M + H] + ions	LLOQ = 5 ng/ml	5–5000 ng/ml	Caffeine and its three primary metabolites in rat plasma		[22]
17	GC-NPD	Stationary phase: capillary fused silica column Mobile phase: carrier gas, helium (1 ml min^{-1})	Detection was made by using nitrogen phosphorus detector LOD = 0.02 μg/ml LOQ = 0.05 μg/ml	0.05–500 μg/ml	Caffeine in teas, coffees, and eight beverages	Caffeine in: Nescafe coffee = 246.8 μg/ml Coffee seed = 267.5 μg/ml Red Bull = 297.9 μg/ml, while other samples contained less caffeine	[23]
18	Infrared spectroscopy	Fourier transform infrared spectroscopy (FT-IR) method	The measurement was done at 1659 cm^{-1} using a baseline established between 1900 and 830 cm^{-1} LOD = 3 mg L^{-1}	NA	Caffeine in roasted coffee samples	Recovery of all samples ranges from 94.4 to 100.1%	[24]

Table 2.
The various analytical methods for the determination of caffeine present in coffee.

Author details

Hemraj Sharma
Department of Pharmacy, Shree Medical and Technical College, Bharatpur, Nepal

*Address all correspondence to: hemrajsharma.hs50@gmail.com

References

[1] Shah, Seth Spiller MA. Textbook of Pharmacognosy and Phytochemistry. 1st Ed. New Delhi: Elsevier, Reed Elsevier India Private Limited; 2010

[2] Spiller MA. The chemical components of coffee. Caffeine. 1998; **1998**:97-161

[3] Wolfrom ML, Plunkett RA, Laver ML. Coffee constituents, carbohydrates of the coffee bean. Journal of Agricultural and Food Chemistry. January 1960;**8**(1):58-65

[4] Belitz HD, Grosch W, Schieberle P. Coffee, tea, cocoa. Food Chemistry. 4th ed. 2009:938-970

[5] Heckman MA, Weil J, De Mejia EG. Caffeine (1, 3, 7-trimethylxanthine) in foods: A comprehensive review on consumption, functionality, safety, and regulatory matters. Journal of Food Science. 2010;**75**(3):R77-R87

[6] Dewick PM. Medicinal Natural Products: A Biosynthetic Approach, 3rd Edition. Chichester, West Sussex, England: John Wiley & Sons, Ltd; 2009. ISBN: 978-0-470-74168-9

[7] Jalal MA, Collin HA. Estimation of caffeine, theophylline and theobromine in plant material. The New Phytologist. 1976;**76**(2):277-281

[8] Sharma H, Sapkota HP, Khan S, Bogati SB, Sapkota B. Estimation of caffeine content on various Brands of tea of Nepal. India and China Consumed in Local Market of Nepal. 2019;**2019**:1-4

[9] Vichare V, Mujgond P, Tambe V, Dhole SN. Simultaneous spectrophotometric determination of paracetamol and caffeine in tablet formulation. International Journal of PharmTech Research. 2010;**2**(4): 2512-2516

[10] Sharma H, Reddy MA, Babu CN, Bhatta HP, Wagle N, Sapkota HP, et al. Method development and validation of dual wavelength UV spectrophotometric method for simultaneous estimation of Paracetamol and caffeine in combined dosage form by internal standard method. Asian Journal of Chemistry. 2015 Jun 15;**27**(12):4666

[11] Gopinandhan TN, Banakar M, Ashwini MS, Basavaraj K. A comparative study on caffeine estimation in coffee samples by different methods. International Journal of Current Research in Chemistry and Pharmaceutical Sciences. 2014;**1**:4-8

[12] Misra H, Mehta D, Mehta BK, Soni M, Jain DC. Study of extraction and HPTLC-UV method for estimation of caffeine in marketed tea (Camellia sinensis) granules. International Journal of Green Pharmacy (IJGP). 2009;**3**(1):47-51

[13] Srdjenovic B, Djordjevic-Milic V, Grujic N, Injac R, Lepojevic Z. Simultaneous HPLC determination of caffeine, theobromine, and theophylline in food, drinks, and herbal products. Journal of Chromatographic Science. 2008;**46**(2):144-149

[14] Babu VS, Patra S, Karanth NG, Kumar MA, Thakur MS. Development of a biosensor for caffeine. Analytica Chimica Acta. 2007;**582**(2):329-334

[15] Mumin A, Akhter KF, Abedin Z, Hossain Z. Determination and characterization of caffeine in tea, coffee and soft drinks by solid phase extraction and high performance liquid chromatography (SPE–HPLC). Malaysian Journal of Chemistry. 2006; **8**(1):045-051

[16] Belay A, Ture K, Redi M, Asfaw A. Measurement of caffeine in coffee beans with UV/Vis spectrometer. Food Chemistry. 2008;**108**(1):310-315

[17] Casal S, Oliveira MB, Ferreira MA. HPLC/diode-array applied to the thermal degradation of trigonelline, nicotinic acid and caffeine in coffee. Food Chemistry. 2000;**68**(4):481-485

[18] Jl B, Tarka S Jr. HPLC determination of caffeine and theobromine in coffee, tea, and instant hot cocoa mixes. Journal of Food Science. 1983;**48**(3):745-747

[19] Perrone D, Donangelo CM, Farah A. Fast simultaneous analysis of caffeine, trigonelline, nicotinic acid and sucrose in coffee by liquid chromatography–mass spectrometry. Food Chemistry. 2008;**110**(4):1030-1035

[20] Amare M, Aklog S. Electrochemical determination of caffeine content in Ethiopian coffee samples using lignin modified glassy carbon electrode. Journal of Analytical Methods in Chemistry. 2017;**2017**:1-8

[21] Ly SY, Jung YS, Kim MH. Kwon Han I, Jung WW, Kim HS. Determination of caffeine using a simple graphite pencil electrode with square-wave anodic stripping voltammetry. Microchimica Acta. 2004; **146**(3–4):207-213

[22] Choi EJ, Bae SH, Park JB, Kwon MJ, Jang SM, Zheng YF, et al. Simultaneous quantification of caffeine and its three primary metabolites in rat plasma by liquid chromatography–tandem mass spectrometry. Food Chemistry. 2013; **141**(3):2735-2742

[23] Sereshti H, Samadi S. A rapid and simple determination of caffeine in teas, coffees and eight beverages. Food Chemistry. 2014;**158**:8-13

[24] Garrigues JM, Bouhsain Z, Garrigues S, de la Guardia M. Fourier transform infrared determination of caffeine in roasted coffee samples. Fresenius' Journal of Analytical Chemistry. 2000;**366**(3):319-322

Section 5

Coffee Commercialization

Chapter 6

Coffee Commercialization in the Bolaven Plateau in the Southern of Lao PDR

Saithong Phommavong, Maliphone Douangphachanh and Khanhpaseuth Svengsucksa

Abstract

Coffee commercialization is a part of agricultural commercialization, which is recently tended to be increasingly integrated into the national, regional, and global market to increase the export earnings. In this context, coffee commercialization is discoursed in pertinent to six perspectives involving agricultural commercialization, the linkage of agriculture and commercialization, new thinking of agricultural commercialization, productive resources for coffee commercialization, process of coffee commercialization, and impact of coffee commercialization. Agricultural commercialization has to reconstruct of relation to food security and gender role in agricultural commercialization. The implication of empirical analysis of FATE research project in the Southern of Lao PDR highlights three novel features of coffee commercialization including productive resources for coffee commercialization, process of coffee commercialization, and impact of coffee commercialization. The primary conditions required for coffee commercialization are productive resources including land, capital, and labor. The coffee commercialization is essentially processing through plantation, nurturing, harvesting, and marketing. The commercialization provides livelihood impact arena for local farmers' food security and employment opportunities including for women and creates compact gender division of labor in different processes of coffee commercialization.

Keywords: commercialization, commercialization and agriculture, commercialization process, coffee production, Lao PDR

1. Introduction

Agriculture commercialization has long debated regarding commercialization and rural living standard [1], commercialization of subsistence agriculture in developing countries [2], process of commercialization to determine nutritional outcomes [3], smallholder agriculture commercialization [4–7], and commercialization and agricultural extension services [8, 9]. Some provide a background of concepts and theories of commercialization that marketable surplus as a concept and neoclassical theory of agricultural commercialization [8, 10]. Other examine case empirical studies of commercialization of agriculture [3, 5, 7, 9, 11]. Most of the empirical study take general cash crop and limited focus on coffee commercialization [12–16].

IntechOpen

Generally, agriculture commercialization is the process of changing the path of putting agricultural production into the market. It is an approach to changing farmers' perspective on farming as a subsistence activity toward a profitable business, if economic sustenance in developing countries is to be achieved ([17], p.172). Other five decisive characteristics of agricultural commercialization apart from profit maximization are production for sales, fulfill the needs and preference of customers, business-oriented, and gaining success [8]. Some highlight agricultural commercialization "as an agricultural transformation process in which farmers shift from mainly consumption-oriented subsistence production toward the market and profit-oriented production systems" [7]. Therefore, agricultural commercialization has become the first priority to boost economy and development by enhancing food security and increasing crop productivity for export. However, commercializing agriculture is a crucial and challenging option when the core question is required to take into consideration of "how farmers can adapt and develop a subsistence farming into a profitable business and how to promote commercial agriculture". Besides this, the concern of food security also requires the commercial transformation of subsistence agriculture.

Agriculture is a strategic sector promoting economic growth, development, and sustainability in most least developed and developing nations including that of Lao PDR. It underpins food security, rural development, and export expenditure, which is the mainstay of economic contribution. Agriculture is also becoming dominant to livelihood activity of smallholder and large-scale farmers. Despite the fact that agriculture production in many least/developing countries tended to be lag behind and experienced a chronic problem causing the rise of poverty and food insecurity. However, with the growing integration of the global market, agriculture production tended to be increasingly integrated into the regional, national, and global market; while agricultural commodity tends to increase the export earnings.

Hence, least developed and developing nations have to confront many difficulties to develop and promote the agriculture sector, both internal and external. The internal difficulties include low productivity, lack of technical support, low skill capacity, poor infrastructure, and deficient institutional. Also, the external difficulties mean highly competitive within the market as the trend of globalization, liberalization, and particularly commercialization are leading to the growth of market integration. Therefore, transforming agriculture outputs from a subsistence economy to market-oriented is an option, called "agriculture commercialization."

Lao People's Democratic Republic or Laos has policy to support agricultural commercialization of cash crop products, including coffee. Lao's agriculture system was subsistence farming in the past. People produced a little surplus for exchange after household consumptions. In the last four decade, the country started agricultural transformation by introducing the New Economic Mechanism strategy (NEM) in the Fourth Party Congress in 1986. This strategy is aiming to commercialize agriculture by increasing cash crop production and encourage smallholder farmer to become integrated in regional and international markets. The transformation of agriculture through the commercialization process has been carried out with a strategy of "an open market economy." Various programs have been introduced by the government promotion of commodity production (PCP) and to support the transformation of commercialization by reviewing a significant issue related to the agriculture system of Lao PDR. Hence, agricultural production has been implemented under the PCP. Therefore, commercial plantations have expanded dramatically, including coffee.

Coffee plantations were first introduced to Laos around 1920s by the French Colonial Empire in the Bolaven Plateau, southern of Laos [18], which is a top commodity for export-led in Laos nowadays [19]. Today, about 99% of Lao coffee is produced in the Bolaven Plateau, approximately 15,000 smallholder households or 69% of farms depend on coffee production as their primary source of income and harvest of green coffee about 28,000 tons in 2013 [20, 21] and with 1053 smallholder family farming produce organic coffee [22]. Coffee producers in southern Laos include private investors, farmers' cooperative, and individual farmers [20, 23]. The total coffee plantation accounts 86,763 ha; a portion of a coffee plantation in Bolaven Plateau is the largest highland area in the country that covers more than 70,000 ha [24]. Therefore, coffee is one of the most outstanding crops cultivated in Laos, mostly by smallholder farmers who live in rural areas, where it is aimed for rural development and sustainability effort. Commercialization of coffee has been promoted by the government and private sector under the NEM in the 1990s, particularly to export the potential markets, at satisfactory market demand, and fair price for farmers. About 95% of the total coffee sold in international markets [24] constitutes coffee commercialization.

This chapter aims at discourse and shed light on some issues of coffee commercialization regarding their concept and theory of agricultural commercialization, nexus of commercialization and agriculture, rethinking of commercialization and agriculture, productive resources for coffee commercialization, process of coffee commercialization, and impact of coffee commercialization. The implication of empirical work of the FATE research project, a collaborative project of five countries: Laos, Nepal, Rwanda, Bolivia, and Switzerland during 2014–2020 is applied as a case study. The focus of the project is the feminization of employment in coffee production at the Bolaven plateau, the southern of Laos. The research project is mainly focused on how women are involved with agricultural commercialization, particularly in coffee commercialization process. The chapter is thus expected to enhance the novelty of knowledge in the field. The chapter, after introduction continues with discourse of literature in the field and follows with the empirical implication of the coffee commercialization, and ends with concluding discussion.

2. Conceptualization of commercialization

Commercialization has different meanings and concepts. Many scholars provide that commercialization is to do with the production side, while others stress the importance of the marketing and distribution channels. Commercialization is thus a relation of the process of production, processing, transportation, and marketing of the product. Commercialization is involving several agencies to fulfill different tasks of production, processing, transportation, and marketing. Taking commercialization as a process, nature and rate of commercialization are concerned with a group of farmers, peasants, whose livelihood is relying on agriculture.

Commercialization is defined by various disciplines. Over time, most literature on commercialization has operated the definition of commercialization in a number of ways; in most cases, commercialization has been found in the concept of development studies relating to agriculture. In common term, "commercialization" refers to the process and the proportion to finalize agricultural production into the market. It is a subset of the broader process of innovation that is driven by market and profit motives, with firms and others seeking to gain a positive return on investment and marketing, including through the creation of competitive niche markets ([25], p.37). Commercialization is thus a process to marketing, particularly high-value cash crops such as horticultural crops and primary food crops such as wheat [6]. However, others view commercialization not only for the marketing, but also commercialization should be profit maximization, response to the needs and interests of buyers, and fulfill the achievements of business [8]. Some scholars focus on the issues of market management to have stable procurement of raw materials and the capacity to access regional and foreign markets [26]. This part deflated the viewpoint that commercialization processes enhance the economy of the local livelihood and business sector. On the one hand, in order to achieve developed and sustainable status, agricultural commercialization has to follow a pathway from subsistence farming to a more profitable business of market-oriented economy and shifting to cash crops cultivation. In that case, commercialization of agriculture is a process of cash crop production which is produced for sale [27].

In this context, the concept of commercialization is set around the framework of agriculture involving the transformation pattern from subsistence-oriented economy to the market-oriented economy by commercializing agricultural cash crop productivity. This also underlined that the transition of smallholder farming from subsistence to more commercialized enterprises is a key feature of agricultural transformation in the process of economic development [28, 29]. Collectively, the agriculture commercialization allows numerous smallholder farmers to participate in the market economy. Following the pattern of agriculture commercialization, this transformation creates different opportunities which may lead to the raising in return of farmer's assets such as land and labor, encouraging women empowerment and an integrated of market. Hence, commercialization is defined in this chapter as shifting from a highly subsistence-oriented of food production for home consumption toward more cultivate of cash crops, more specialized production and targeting both domestic and international markets.

Viewing the process of commercialization, various scholars highlight of different dimensions. Some researchers specify that process of commercialization is defined by three levels, including subsistence systems, semi-commercial systems, and commercial systems [6, 17]. These processes may be defined as a mean of generating more income into the household by leaning the advantages of subsistence production. Commercialization process is certain to modernize agriculture by increasingly using the complexity of the production process, and utilizing new technology and mechanization in the production process [4]. Processes in smallholder commercialization are also seen as a pathway of non-agricultural sectors integration to overall economic structure [13]. In that sense, various agencies, both public and private institutions, involve with commercialization at different magnitudes and contribute to the success and failure of commercialization. Some of the factors contributing to the success and failure of the commercialization are, for instance, (1) effective institution, (2) improved infrastructure, (3) knowledge management, (4) adequate incentives, (5) stakeholders' initiative, and (6) a conducive environment [30]. For the achievement of the commercialization of agriculture, it has to take all factors into account as such.

3. Nexus of commercialization and agriculture

What is the relation between commercialization and agriculture? The relation of the commercialization and agriculture is involving market surplus, productive resource, and value chain. A relation of commercialization and agriculture can be seen through the market surplus. The market surplus of agricultural produce is an indicator of commercialization measurement. The greater market orientation of the producers means commercialization, whereas small proportions of surpluses are more subsistence-oriented [8]. Besides, Bandara [8] define that the commercialization of agriculture is supplying of agricultural production to market which target high profit ratio and ensuring that meet the need of consumers' interests. This type of business is called "agri-business" that use business management concept to help farmers'success in their business ([8]: 6).

Goletti et al. [26] stated that agricultural commercialization is a complex process that comprising various dimension (e.g. famers, technology, market and marketing, finance, institution, infrastructure, consumers). Also, the transition from subsistence to different degrees of commercialize (e.g. low, medium, high, and advance) leads in different value. They highlight that it is possible to measure agriculture commercialization and suggested that to promote commercial agriculture should create applicable systematic of commercialization [26].

Mahaliyanaarachchi [31] also stated that agricultural extension is an on-going and informal educational process that takes place for a long time. Also, agricultural extension's benefits cause farmers' livelihood development. Therefore, it is recommended that it needs to improve farmers' knowledge, skills and influence their new way of thinking in agriculture activity such as apply new technologies, farming activities, and marketing ([8, 31]: 14).

Rivera [32] mentioned that agricultural extension is also primary element that impact the degree of commercialization of agriculture. Nowadays, agricultural extension is a primary agricultural product with a stable price (commodification of agricultural extension and the transforming of knowledge into a product for sale) supported transforming both public sector extension and private business sector on technology transfer in agriculture sector [8, 32]. Concepts of commercial agriculture extension can be observed in three dimensions. Firstly, the agricultural extension is observed as a commercial product or service that exchanged by two parties. First party is extension providers as sellers and other party (farmers) as buyers. Secondly, it is applying primary concept as supply and demand. Hence, agricultural extension become a demand-oriented activity. Finally, extension can be viewed as inputs (e.g. fertilizer, hybrid seeds, agro-chemicals, machinery, and others), which is fundamental for the commercially oriented farming [8].

Some studies discovered that most developing countries in Asian and Latin American used the land for cash crop production since the 1980s [2, 3] because of encourage to increase cash crop export and local income improvement [2]. At the same time, a study of Omiti et al. [4] found that agricultural commercialization in Kenya has a constraint in creating a value chain. Various crops take different forms of market access such as maize has inadequate market access, while tomatoes, milk, and kales production have good market access. In short, high market integration and reasonable market access support diversification into high-value mixed enterprises [4]. One research on an impact of cooperatives on smallholders' commercialization behavior revealed that smaller farmers tend to supply less product because they target to increase the price while larger farmers maximize their farm product [33]. Jaleta et al. [13] pointed out that smallholder commercialization concept goes beyond the marketing of surplus staple products because it consists of household input use decisions, major objectives of production, household participation in input and output markets, degree of specialization in production, and dependence on markets for income and consumption. It is a highlight to emphasize on smallholder commercialization level [13].

Based on Zhou et al. [7]'s study in the southern africa who pointed out that agricultual commercialization produces both positive and negative effects. The authors pointed out some key findings that agricultural commercialization produces both positive and negative effects. The positive effects at the household level reflected from increased productivity, family employment, more household income through market participation and employment, a better consumption diversity, nutritional welfare development, improved education, health and welfare, and household living conditions development [7, 34]. Besides, at the society arena, commercialization contributes to food security, poverty alleviation, rural and urban employment creation, improved livelihoods and social status, and economic growth via productivity and investment [7, 34]. The commercialized leads to negative effects such as no improvement household nutrition and livelihoods of the poorest, more risk complex market, household food insecurity, and insufficiency food [7, 17, 34]. Besides, the commercialization influences to income inequalities [7, 17]. Agricultural commercialization among poor smallholders is an issue that requires to pay attention to reduce poverty, improve household food and nutrition security, and foster growth in rural areas [35].

4. Rethinking of agricultural commercialization

Various discourses on agricultural commercialization provide background concept and the relation of the issue with regard to the definition of terms, process, and their relations. In this part, the interest is to reconstruct the agricultural commercialization to rethink how commercialization will lead to food security and gender role in agricultural commercialization.

A classical thought argues that the commercialization of agriculture is a cause of poor nutrition [3]. Others argue that subsistence agriculture may not be a viable activity to secure sustainable household food security [13, 36]. Agricultural commercialization is thus provision of both opportunities for food security and income earning. One of the pathways to food security is to promote food crop productivity [5] that is to enhance household food supply. Pathways to promote food crop productivity are improving access to credits and inputs, input cost management, high-value crop investment, and investment in infrastructure and human capital. The element of food security is related to accessibility, availability, stability, and utilization of food [4]. Measuring food security can apply concept of consumption or calorie intake method, expenditure, nutritional food sufficiency, coping strategies, and resources availability [37].

The most crucial question is to what extent the agricultural commercialization contributes to food security. The effect of agriculture commercialization on income and productivity seems positive, whether on the household level or for large scale producers; however, it depends on the allocation [2]. The allocation refers to resources management including land, labor, and capital toward production activities, which come to the decision of how to allocate or distribute these resources in order to achieve a greater benefit from this transformation. Since the household food availability and consumption depended on agriculture in many least developed and developing nations; the impact of agriculture commercialization should be evaluated for better application of the conceptual framework of agriculture commercialization. Despite the fact that the process of agriculture commercialization is positively affected, the income gains from selling crop in the market, which also enhancing household capacity to effort for food. However, food security and income have not yet shown a direct relationship. In terms of food security, there are still arguments for and in opposition to smallholder commercialization as a pathway for making sure household food security. Food security is also determined by farming activities, in which the shifting to cash crop may be led to the decrease of food crop cultivation. On the one hand, smallholder commercialization is assumed to have damaging outcomes on household dietary and food safety status.

Indeed, the adverse impact of the process of commercialization on food security is still debatable among scholars. For example, Von Braun [38] argued that "commercialization has a direct effect on household's earning degree which likely results in a rise in food and non-meals expenditure" ([38]: 187). On the other hand, the famous Engel's Law claimed that there is an inverse relationship between the proportion of food expenditure and overall earnings, which people are willing to spend more on food as their incomes increased [39]. Therefore, food security pushed further concern on the impact of agriculture commercialization; however, the correlation between the raise of income earning and nutrition can also make a positive difference.

The second issue that needs to rethink of the agricultural commercialization is gender role in agricultural commercialization. Some scholars attempt to provide measurement and effect of commercialization of agriculture. Different degrees of agricultural commercialization are well defined. Primary characters of agricultural commercialization are farmers, technology, market and marketing, finance, institution, infrastructure, and consumers [26]. Among them, a group of factors that lead to project success are commercialization that needs involvement a large number of women [26]. However, different participation of stakeholders' involvement in agricultural commercialization leads to different magnitudes of agricultural commercialization. The high and advanced dimension of agricultural commercialization need lower involvement of farmers and need industrialization of agriculture, for instance.

Gender role is a decisive component to agricultural commercialization, particularly to commercialize of agriculture at the micro-level (household and company levels) that commercialize crop improves local livelihood. Women play crucial role in both commercialization of agricultural produce and secure household food supply. A study of Sørensen [40] found that commercial of food crop constructed a new gender relation. Men play role dominantly in food production which reflects of patriarchal practice. Also, the socio-economics lead to different bargaining power. Women from a better economic have better bargaining power when compare with other women who have not good economic conditions [40]. Gender power relation within households provides the benefits arena to women. It is expected that women gain cash income from commercialization, but whether or not women benefit depends on how decisions are made within the households [26]. In fact, labor wage is not equally distributed for women and men for the same amount of work. Instead, women are facing commercialization of agricultural work plus the burden of household chores. The benefit of smallholder commercialization on the gender dimension also depends on a specific commodity when gender-labor demand and on the decision to the income generated [13]. A confirmation is that commercialization of agriculture provides disadvantage for women because of gender inequality in access to productive resources unless they become membership of farmer groups [41].

5. Productive resources for coffee commercialization

5.1 Land acquisition for coffee production

The most fundamental factor in agricultural production is land acquisition. The government of Lao PDR has the vision to promote the agricultural sector; the priority is to specify the existence of cultivated land expansion in the national framework. The survey and allocation of agricultural land have been carried out nationwide to allocate land to the district level. Land titles have been issued for farmers at the village level, equivalents to 43.2% of all villages across the country and cover 37.1% of the total districts [19] or about 800,000 pieces (Open Development Laos, 2019).

Land classification and land title are required to take into account as the factor affecting agriculture production because the potential of arable land, particularly agriculture land, refers to the areas that can be brought for cultivation with soil, water, and climatic suitability. Pingali and Rosegrant [17] noted that "Agricultural commercialization means more than the marketing of agricultural output; it means [that] the product choice [s] and input use decisions are based on the principles of profit maximization" (n.p). Therefore, the land is a determinant factor of agriculture product inputs, acquisition of use or ownership rights to large areas of land for production of agricultural commodities, by farmers has recently attracted considerable interest.

Land use for coffee commercialization at the Bolaven Plateau, southern Laos consists of two forms, local own, and foreign private own. The former form is mainly smallholder coffee commercialization, who are local people, in which land use is only approximately 1–2 ha for coffee plantation. Based on the FATE household survey in 2015, the average land use for smallholder who could commercialize their coffee is 2.64 ha with a minimum of 0.10 ha, and the maximum land size is 20 ha. The majority of the farmers, about 45% owned land of 2.0–4.9 ha of agriculture land and 51% of these coffee land is used to plant Arabica coffee. There shows correlation between coffee commercialization and land use at a significant level. Access to cultivated land by the smallholders in the plateau is by three approaches including *Chap Chong* (Lao word means free land acquisition), land inheritance, and purchasing land use right [42].

The latter form, land use for coffee plantation is by foreign private ownership. After the economic liberalization policy has been applied under the NEM, the government of Laos (GoL) gives the permit to foreign investors to allow for land concession to various industries also for coffee plantation. The rapid proliferation of land concessions has been granted by the GoL to investors who are seeking to capitalize on the plateau's agriculture, forestry, hydropower, and mineral commodity chain potentialities [43–45]. The first agriculture land concession for a foreign private company to coffee planter was granted to Asia Tech in 1991 about 12,000 ha. A plenty number of smallholder coffee producer turned into the host of coffee concession, 37 of 84 villages in the Bolaven Plateau hosted at least one coffee concession in the administrative village; while 10 villages hosted two or more concession projects [46]. Today, the idea of land concession is still debatable in the institutional level; whereas, agriculture land concession demonstrates primarily to coffee commercialization.

5.2 Capital formation

Capital input is another significant productive resource of coffee commercialization in which farmers have to have some capital to start farming as a running cost such as for planting, nurturing, processing, and marketing. In order to process all of these farming activities, farmers have to accumulate the capital from various sources and spend for those activities. Smallholder coffee producers seek their ways to find the capital to support their coffee production. Access to capital can be approached in several ways, including through rural development fund, banking institutions, private money lenders, relatives, and friends [42]. However, the study found that the production of coffee holds an average cost of production, representing as fixed and variable costs. Therefore, capital input in this context will consider all the cost of production, both fixed and variable costs. The study showed that the production of coffee cost about 9.13 million kips per hectare, 0.04 million kips per hectare as a fixed cost, and 9.09 million kips as variable cost. For the variable cost, tools and equipment using in the process of coffee production are including, which take about 19.59% of the total cost of production. Petroleum gas cost the highest with an average of 19.55% of the total cost of production; while the cost of wage labor was only about 22.57%, fertilizers cost about 22.47% of the total cost of production, and other administrative costs about 15.82%.

5.3 Labor input for coffee production

In demographic change, the total population of Lao was only 7.13 million people, and 53.93% are below the age of 25, in 2017 [19]; however, the core of the rate of contribution in labor force still considers low. The agriculture sector contributed 62.46% of the GDP growth in 1990 and started to decline in 2017, and recently contributed only 18.55% of the total GDP growth rate (Lao Statistic [47]). While the number of labors participating in agriculture has been fell from 71.3% in 2011 to 65.2% in 2015 [48]. Therefore, the major paradigm of labor input in relations to agriculture development came into attention, specifically in coffee production because it increased from 52.01 tons in 2011 to 99.78 tons in 2015 [19], which means labor demand for coffee production is declining.

Hence, labor as a mean of productivity, engaging in every process of production shows relatively correlation with one another, including surplus and earning. Traditionally, the use of household labor in farming production is important as the part of labor market and production input as well. The study noted that in general, the rate of self-employed farming took about 91% of the total category of occupation; where was correlated to the number of lands owned.

Hence, there is no surprise to the pattern of employment where the wage of self-employed (per month) took the second place of highest salary after the wage salary (per month). It is possible to conclude that working in the coffee farming tends to reach a higher sense of stability; wage labor becomes more specialized and gained expertise through working experiences from years to years as wage labor already carries some fundamental skill working in agriculture or coffee planting. In the past, people always exchanged one another with labor, weed and harvested coffee.

To focus on labor input in the coffee production community, first understand the pattern of labor mobility. During the process of coffee plantation, nurturing and harvesting process are the most significant steps that requires quite numerous of labor; therefore, coffee producers required to hire wage labors to work in the coffee garden all year round particularly in production seasons. As a result, wage labor in coffee production has been shifted around during seasonal and non-seasonal periods.

Labor demand is determined by the cultivation land or the size of the farm. Coffee production is labor intensive; thus, labor is required in different process of coffee plantation. Hence, the greater number of labor results in increasing productivity; similarly the minimum number of labors used in the nurturing process also leads to a decrease in productivity. Therefore, in the process of coffee production, both household and wage labor represent a complex pattern, this also includes the number of women participating in this coffee community as well.

The pattern of employment in the coffee farm at the Bolaven Plateau seemed flexible in term of hiring pattern. There show four patterns of employment including wage employment, non-farm self-employment, a permanent worker for farm, and seasonal worker for the farm. Wage employment takes major account of the off-farm employment (56.4%), where local people still contribute to the employment in different occupation mainly agricultural work, public sector, and private work. Seasonal farm employment shares a larger account than permanent work for the farm. The major work for seasonal employment is an agricultural worker (general/clearing weeds) accounted for 28.6% and harvest only about 17% [49].

Although, there is a variety of labor employment pattern that shifted within the coffee production at the Bolaven Plateau; a significant factor affecting labor mobility is the cycle of the temporary and permanent worker during the farming season. The temporary worker based on payment as known for wage employment usually increased in the farming season. Whereas, permanent worker based on types of work, in a small farm producer or a large company is small proportion. Therefore, the rate of permanent employment relied on the size of the land (for small farm producer) and the payment or benefits (for the large company producer).

Seasonal farmworkers were mainly migrated from nearby provinces, districts, and villages who come for different farm works including coffee tree plantation, fertilizing, clearing, and harvesting. The major account of the farm employment was general/clearing weeds (28.2%) and followed by harvest only (16.9%) [50]. However, many types of work are required to be done by wage labors from the same village, including planting seed and weeding, whereas some processes are required to be done by wage labor migrated from nearby village or district, specifically during the harvesting season. Furthermore, during the harvest season, the rate of wage labor slightly increased because labor demand is high during the peak season.

6. Process of coffee commercialization

6.1 Coffee plantation

The plantation is the primary process of coffee commercialization that provides various methods to grow coffee. There are three types of coffee plantation including planting with seeds, planting to cut the top, and planting to prune and every method depends on the size of seeds depends on the geographical area, the elevation, and the climate of the planting areas. The first method of coffee plantation, planting with seeds is applied with a limited of land, this technique is widely adapted because it takes only a year to growth, with a simple of nurturing and fertilizing. Hence, this technique concerns different preparation step before the plantation, including soil, seed, and fertilizer. In many cases, farmers mixed between soil and manure as a basement of the plant.

Planting to prune is another method mostly used in the lowland areas with dry weather. In most case, the coffee plant can be easily damaged. For the small coffee tree, it will grow only if the branches break to the top as needed. Nurturing is quite tricky for this planting method due to pests and weeds. However, if the coffee tree gains a bigger size with great roots, soil, and watering condition, the branches can grow easily. The third coffee planting method, the planting to cut the top is suitable for the planting areas located on high altitude over 900 m above sea level,, due to the plateau areas cover with less sunlight and benefit from rains for watering condition. Arabica is the most popular beans growing in the plateau areas; however, the cutting or pruning technique requires expertise and experience because this method required to prune only one branch at the top. This method seems to be mostly used in the Bolaven Plateau when the geography is profitable. Nurturing and fertilizing are simple for this planting method.

6.2 Nurturing

The process of nurturing coffee plant depends on the planting method, where planting to prune is the most difficult for nurturing due to pest and weeds. The common nurturing techniques are related to six steps including planting season, watering condition, coffee stem covering, weeds and prevention, fertilizer, and coffee pruning. The first step is to select planting season. The best season for growing coffee plant is rainy season starting from May to June in order to reduce watering condition for seedlings. However, without the rains, it is required to water the plants about 10 days before covering the plant with dry straw for maintaining soil moisture, which is labor-intensive and time-consuming. The next step is watering of the tree. Watering the coffee plant should not be overwatering. The new coffee plant is required a limited amount of water to help for sitting up; therefore, planting coffee during rainy season is the best method for nurturing. Without watering, the coffee plant can be easily damaged because nutrition relies on water to dissolve. The study found that farmers only relies on rainfall for watering condition. Coffee stem covering: there are many types of materials used for stem covering, including Napier grass. When Napier grass damaged, it produced a large amount of nitrogen, which is useful for coffee. However, in Laos, straw is widely used, replacing napier grass to reduce the cost as well. Covering the stem, a third step, of the coffee plant should be 10–20 cm away from the plant for preventing some pests or insects. With 1-m width and greater than 10 cm of thickness will help for maintaining the soil moisture. The study found that the limitation of using stem covering, as farmer only used dry straw and grass, and some areas also experienced less rainfall in a year (6–8 months or less) as a result of lowering the productivity.

The weeds and prevention, a fourth step of coffee plantation is to get rid of weeds, farmers are sometime using shovel, knife, or cutting. However, the most used to deal with this problem is chemical or pesticide with the exception of the coffee farm of coffee producer cooperative (CPC) which has strict control of chemical to meet the criteria of the organic certified product. Diuron is one type of pest control, which farmers used only 200 g for 400 m^2 of cultivated land. Another option for the farmer is pesticide spray, where farmer has to spray 50 cm away from the plant in order to save the coffee plant. The study showed that farmers, particularly the CPC members are more likely to use natural method such as knife and shovel to deal with weeds and pest instead of spraying pesticide or chemical. Household labors are more likely to engage in this process more than another process. Fertilizer is one of the most important step of coffee nurturing. To fertilize the coffee plant, it is required to make the hole around the coffee plant about 5 cm in depth and mainly use the radius of the canopy of the plant and cover the soil with dry straw. Fertilizing in the first to the third year will be much effective. Manure is the best fertilizer used with the proportion of 100–350 grams per time, fertilize three times a year during these consecutive years. Thus, the study noted that 42.46% of coffee producer used organic fertilizer, the coffee shell, and manure. On the other hand, the other 57.54% do not use organic fertilizer at all because they take only red cherry beans [50].

Coffee pruning is a last important step of coffee nurturing. The productivity of coffee depends on pruning. Without pruning step, the coffee plant can produce a large number of beans, but the year after the production and quality will be decreased because the roots, stem, and branches are already used to produce the seeds. The study recognized that most coffee producers avoid the pruning step as a result of unstable productivity, plenty of older coffee plant were left behind, some dried, and damaged.

6.3 Harvesting

Harvesting is an important process of coffee commercialization to add value to the coffee product and preparation for commercialization. As usual, the coffee cherries will begin to ripen in about November until April. To collect or harvest the

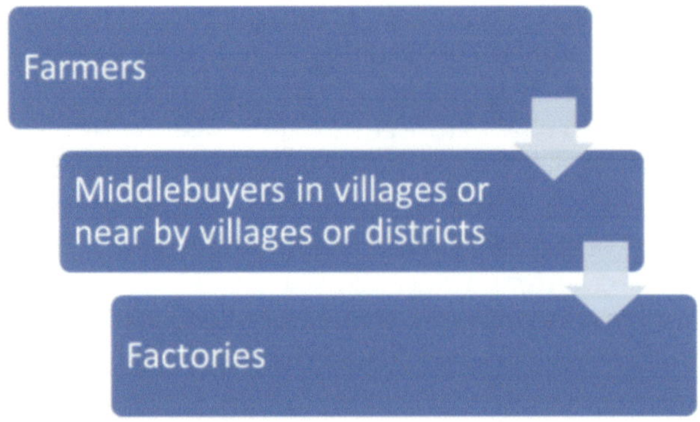

Figure 1.
Coffee marketing process (Illustrated by authors).

best quality of coffee beans, it is to collect the ripened beans in the shade of orange to red. Green coffee beans should be avoided due to poor quality. By using large baskets tidying on the waist or hanged to the neck before, store it into the sack. This step is time-consuming, collecting coffee beans will need to collecting continuously about 3–4 times with a distance of about 20 days because the beans are ripened differently.

Post-harvest processing prepares important step to marketing. Most of the coffee beans particularly Arabica are sold in red bean after picking at the farmgate by the farmers. Processing of the red bean to dried one need facility including machine, that means another capital investment for this facility. The farmers who cannot afford this technique, thus prefers to sell red been. The CPC member groups have the opportunities to process the red bean before marketing at the processing houses. The processing of the red bean provides important procedure for the farmers to involve with difference commercialization process and have higher value added.

6.4 Marketing

Marketing of coffee product highlights the most decisive stage of coffee commercialization during post-harvest. The coffee farmers market their coffee product in various forms, including selling of red cherries, dried beans, milled beans, and roasted beans. Selling the production in the form of red cherry beans is most popular one, where the price is varied between 1500 and 3200 kip per kg. Some groups of farmers who have Robusta and Excellsa coffee prefer dried beans where they can keep for long time and expect to higher value. Recently, some of the farmers also roast their products to serve the tourists locally to get higher coffee value. The marketing circle starts with the farmers and continue to a middleman, and then exporter or factory (**Figure 1**). The middle man will come to purchase coffee beans at the farm gate before processing and wholesaling to the coffee industry such as Dao Heuang coffee factory in Champasak province.

7. Impact of coffee commercialization

7.1 Employment opportunities in coffee production

Coffee production provides main source of employment opportunities for local people. The employment opportunities take four categories including daily wage, monthly, payment under contractor, and payment by product. Farmers normally get daily wage for harvesting, weeding, and applying fertilizer activities. Monthly paid on the other hand only for harvesting task, and payment under contractor is applied for weeding work. In addition, farmers get daily wage paid for harvesting based on product of coffee bean' s weight, where they could get 1 kg of red cherry is about 800–1000 kips. Laborers prefer to receive payment based on the weight because they earn better than a fixed wage which daily fixed-wage is only 30,000–50,000 kips per person. A laborer can pick about 100 or 200 kg, which is approximately 100,000–200,000 kips per day. The payment is higher and competitive during the harvesting. The extra laborer is needed for harvesting and pay by weight during booming of red cherry coffee. Besides, some farmers pay monthly wage only for harvesting, especially in Sedkhod, Phorkhem, and Dong villages under the payment rate of between 700,000 and 1,000,000 kips.

In short, coffee cultivation creates employment opportunities and address seasonal unemployment issue for people in the village, nearby villages as well as people from other districts or provinces, especially during harvesting, weeding, and apply fertilizer in different seasons. It is evident that coffee provide direct jobs to local people and a complementary source of income for them. In fact, the coffee production absorbs seasonal unemployment laborers of agricultural sector, where the farmers can work during the post-harvest season of agriculture.

7.2 Women's participation in the process of coffee production

Coffee commercialization is gender-related working culture that required women participation in farm production in particular work positions including seedling, weeding, fertilizing, harvesting, and processing. Women take most part of the seedling work including prepare seed box and nurturing. Weeding of coffee trees by using traditional tools such as knife to clear grassed out of farm is also carried out by women, while men use machine grass cutter. The coffee cultivation activity requires weeding several times throughout a year which provide employment opportunities and wage earning for women. Women also take part in another activity of the farm, that is fertilizer application. Women participate in harvesting, particularly for Arabica Cartimor coffee which requires soft hands to take care of the young cherries while picking. Women are good at these techniques. The coffee cherry of Arabica variety is not ripened at the single time. Therefore, only red ones are allowed to be plucked. During the booming of the red coffee, cherry need to be picked; otherwise, the red cherry falls and wastes.

Although, women and men spend similar time-use in coffee cultivation together, some tasks of coffee cultivation are gender-based. The tasks require energy (e.g. weeding by using the machine, fertilizing, pruning, washing, and heavy lifting) are men's task, while women take the time-consuming and light work such as harvesting, manual weeding, and sun drying. Furthermore, labor hired on coffee cultivation reflects gender role. There is more female labor than male labor on both manual weeding and harvesting tasks. The harvesting and manual weeding are considered as time-consuming work and tedious tasks. Men prefer to do a task that wastes energy in a short time.

The gender and seasonal employment reveal that gender of hired labor is significantly correlated with employment activity including wage, work hour, benefit, working day, and worker with children at work at the 0.05 and 0.01 level. Female workers also mainly involve in the processing factory for the task of filling and packaging coffee sack. In conclusion, women's role in coffee commercialization is involved with the household or company production levels, particularly for the light and time-consuming works (**Figure 2**).

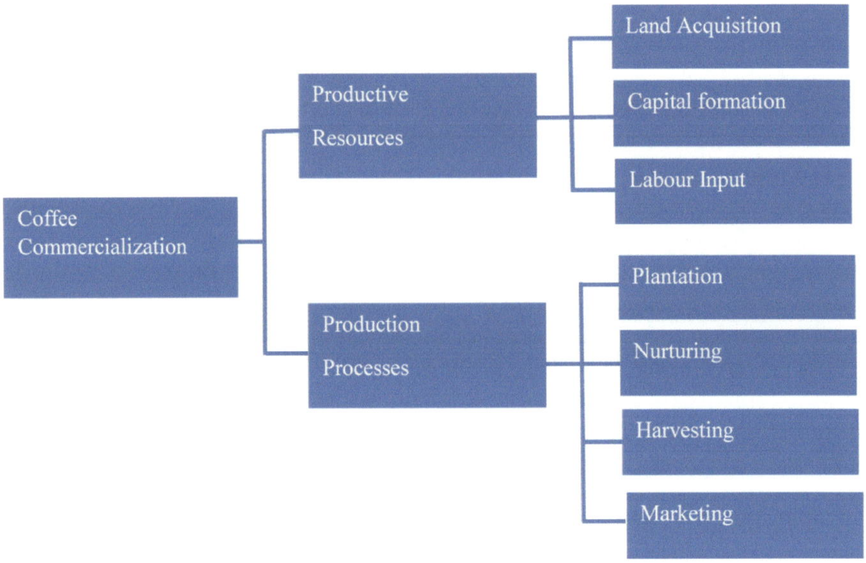

Figure 2.
Coffee commercialization (illustrated by authors).

8. Concluding discussion

Various grounds of literature review provide agricultural commercialization in different aspect, not least commercialization is shifting from subsistence mode of production toward more market-based cultivation of cash crops, more specialized production, and targeting both domestic and international markets. There shows of the nexus of commercialization and agriculture which aims at sales, profit maximization, and meet the needs of customers [8]. To satisfy those conditions, the agriculture needs to be market surplus, productive resources, and value chain added [41]. Agricultural commercialization has to rethink and reconstruct how to meet the need of both food security [5, 13, 36]and gender role in agricultural commercialization [13, 26, 40]. A challenge is to create most favorable conditions for agricultural commercialization to meet the need of food security.

Agricultural commercialization of coffee production highlight factors that plays decisive role in the commercialization processes. Productive resources include land acquisition, capital formation, and labor input constitutes the fundamental elements of the process. Land acquisition including ownership, usage, and benefit form the background of the commercialization. Without these resources, the process might not be fulfilled. Thanks for the farmers who could take forms of land uses such as *Chapchong*, inheritance, and purchase the right to use [42]. Capital and labor also contributes significant impact and direction on commercialization in which they could facilitate and secure the process.

In this context, coffee commercialization process takes form of plantation, nurturing, harvesting, and marketing. Plantation needs detail support of techniques on how to apply to different contexts of coffee plantation including soil, seed, and fertilizer. Nurturing requires special treatment in order to get fruitful harvest. Harvesting is marking as one of the process to indicate of the success of the farm. If the yield is fruitful, this marks as a sign of production surplus for home consumption to secure food supply and ready for-profit maximization. Finally, marketing is crucial indicator of the commercialization, of how the product is selling and how they can maximize profit.

This chapter attempts to highlight some dimensions of gender working culture, particularly women participation in seedling, weeding, fertilizing, harvesting, and processing. In general, it is assumed that women have equal role, time use, and benefit from work. In fact, to create gender equality is not an easy task. The most important function is to share and distribute the benefit from coffee commercialization. Gender role on commercialization is thus needed to equally distribute the task and benefit from the process. On the one hand, the commercialization has to maximize profit as much as possible to women who are considered as the most effective household financial manager. Therefore, the commercialization of coffee is a complex set of factors, agencies, and mediators to facilitate the process.

Author details

Saithong Phommavong[1*], Maliphone Douangphachanh[1,2] and Khanhpaseuth Svengsucksa[3]

1 Faculty of Social Sciences, National University of Laos, Vientiane, Lao PDR

2 Gender Studies Department, Faculty of Art and Social Sciences, Universiti Malaya, Kuala Lumpur, Malaysia

3 International Development Program, National University of Laos, Lao PDR

*Address all correspondence to: sai7512@yahoo.com

References

[1] Reinhardt N. Commercialization of agriculture and rural living standards: El Palmar, Colombia, 1960-1979. The Journal of Economic History. 1983;**43**(1):251-259. DOI: 10.1017/S0022050700029247

[2] Von Braun J, Kennedy E. Commercialization of subsistence agriculture: Income and nutritional effects in developing countries. In: Working Papers on Commercialization of Agriculture and Nutrition. Washington, D.C.: International Food Policy Research Institute; 1986. 79p

[3] Howarth E, James L. Agricultural Commercialization , Nutrition, and the Rural Poor. London: Lynne Rienner Publishers, Inc.; 1990

[4] Omiti J, Otieno D, Nyanamba T, Cullough EM. Factors influencing the intensity of market participation by smallholders farmers: A case study of rural and perri-urban areas of Kenya. Afjare. 2009;**3**(1):57-82

[5] Govereh J, Jayne TS, Nyoro J. Smallholder Commercialization, Integrated Markets and Food Crop Productivity: Cross-country Evidence in Eastern and Southern Africa. Department of Agricultural Economics and the Department of Economics, Michigan State University; 1999

[6] Hagos A, Geta E. Review on small holders agriculture commercialization in Ethiopia: What are the driving factors to focused on? Journal of Development and Agricultural Economics. 2016;**8**(4):65-76. DOI: 10.5897/JDAE2016.0718

[7] Zhou S, Minde IJ, Mtigwe B. Smallholderagriculturalcommercialization for income growth and poverty alleviation in southern Africa: A review. 2013;**8**(22):2599-2608. DOI: 10.5897/AJAR11.1040

[8] Bandara RMAS. Commercialization of agriculture and role of agricultural extension. African Journal of Agricultural Research. 2006;**6**(1):13-22

[9] Gebremedhin B, Hoekstra D, Tegegne A. Commercialization of Ethiopian agriculture: Extension service from input supplier to knowledge broker and facilitator. In: Ethiopian Farmers Project Working Paper 1. Nairobi, Kenya: International Livestock Research Institute; 2006

[10] Tipraqsa P, Schreinemachers P. Agricultural commercialization of Karen Hill tribes in northern Thailand. Agricultural Economics. 2009;**40**:43-53. DOI: 10.1111/j.1574-0862.2008.00343.x

[11] Huggins C. Seeing Like a Neoliberal state? Authoritarian High Modernism, Commercialization and Governmentality in Rwanda's Agricultural Reform [Doctoral thesis]. Ottawa, Ontario: Carleton University; 2013

[12] Dorsey B. And commercial production among smallholder coffee growers in central Kenya*. Economic Geography. 1999;**75**(2):178-195. Retrieved from: http://www.jstor.org/stable/144250%0Ahttp://about.jstor.org/terms

[13] Jaleta M, Gebremedhin B, Hoekstra D. Smallholder commer-cialization: Processes, determinants and impact. In: Discussion Paper of Ethiopian Farmers Improving Market Opportunities. International Livestock Research Institute; 2009

[14] Nestel D. Coffee in Mexico: International market, agricultural landscape and ecology. Ecological Economics. 1995;**15**(2):165-178. DOI: 10.1016/0921-8009(95)00041-0

[15] Simpson T. The commercialization of Macau's cafés. Ethnography.

2008;**9**(2):197-234. DOI: 10.1177/1466138108089468

[16] Taylor PL, Murray DL, Raynolds LT. Keeping trade fair: Governance challenges in the fair trade coffee initiative. Sustainable Development. 2005;**13**(3):199-208. DOI: 10.1002/sd.278

[17] PingaliPL,RosegrantMW.Agricultural commercialization and diversification: Processes and policies. Food Policy. 1995;**20**(3):171-185. DOI: 10.1016/0306-9192(95)00012-4

[18] Lao Coffee Association. 2019. History. Retrived from: http://www.laocoffeeassociation.org/index.php/en/about-lao-coffee/history

[19] CIA. Laos Demographics Profile 2018. Mundi: Index; 2018. Retrieved from: https://www.indexmundi.com/laos/demographics_profile.html

[20] Schönweger O, Messerli P. Land acquisition, investment, and development in the Lao coffee sector: Successes and failures. Critical Asian Studies. 2015;**47**(1):94-122. DOI: 10.1080/14672715.2015.997095

[21] Tulet JC. Coffee's conquest of the tropics. Etudes Rurales. 2007;**180**(2):49-68

[22] Coffee Producers Cooperative. *Bolaven Plateau* Coffee Producers Cooperative. Champasak Province, Lao PDR: CPC Office; 2019

[23] Galindo J, Sallée B, Manivong P, Mahavong P, David A, Homevongsa V, et al. Participative analysis of coffee supply chain in Lao PDR. In: Programme of Capitalization in Support of Rural Development Policy Bolaven Application Point. Ministry of Agriculture and Forestry; 2007

[24] Epprecht M, Weber A-K, Bernhard R, Keoka K, Saphangthong T, Manivong V, et al. Atlas of Agriculture in the Lao PDR: Patterns and Trends between 1990 and. Switzerland: Centre for Development and Environment (CDE), University of Bern, and Ministry of Agriculture and Forestry (MAF), LaoPDR, with Bern Open Publishing (BOP); 2018. p. 2011

[25] Australian Government. 2003. Mapping Australian Science and Innovation: Main Report. 37p

[26] Goletti F, Purcell T, Smith D. ACI Discussion Paper Series DEVELOPMENT. 2003. 19p

[27] Babu SC, Gajanan SN, Sanyal P. Effects of commercialization of agriculture (shift from traditional crop to cash crop) on food consumption and nutrition—Application of chi-square statistic. Food Security, Poverty and Nutrition Policy Analysis. 2014:63-91. DOI: 10.1016/b978-0-12-405864-4.00003-x

[28] Johnston BF, Mellor JW. The role of agriculture in economic development. American Economic Review. 1961;**51**(4):566-593

[29] Johnston BF. Agriculture and structural transformation in developing countries: A survey of research. Journal of Economic Literature. 1970;**8**(2):369-404

[30] Brandt L. ACI Discussion Paper Series. Agrifood Consulting International; 2016

[31] Mahaliyanaarachchi RP. Basic of Agricultural Extension. Colombo: Godage International Publishers; 2003

[32] Rivera WM. The changing nature of agricultural information and the conflictive global developments shaping extension. Journal of Agricultural Education and Extension (Wageninen). 2000;**7**(1):31-41

[33] Bernard T, Taffesse AS, Gabre-Madhin E. Impact of cooperatives on smallholders' commercialization behavior: Evidence from Ethiopia. Agricultural Economics. 2008;**39**(2):147-161. DOI: 10.1111/j.1574-0862.2008.00324.x

[34] Von Braun J, Kennedy F. Agricultural Commecialization, Economic Development and Nutrition. Baltimore, Maryland: IFPRI, Johns Hopkins University Press; 1994

[35] Carletto C, Corral P, Guelfi A. Agricultural commercialization and nutrition revisited : Empirical evidence from three African countries. Food Policy. 2017;**67**:106-118. DOI: 10.1016/j.foodpol.2016.09.020

[36] Pingali P. From subsistence to commercial production system: The transformation of Asian agriculture. American Journal of Agricultural Economics. 1997;**79**(2):628-634

[37] Maxwell D, Wiebe K. Land Tenure and Food Security: A Review of Concepts, Evidence, and Methods. Land Tenure Center Research Center, University of Wisconsin Madison; 1998

[38] Von Braun J. Agricultural commercialization: Impacts on income and nutrition and implications for policy. Food Policy. 1995;**20**(3):187-202. DOI: 10.1016/0306-9192(95)00013-5

[39] FAO. The State of Food Insecurity in the World—High Food Prices and Food Security—Threats and Opportunities. Rome, Italy: Food and Agricultural Organization of the United Nations; 2008

[40] Sorensen P. Commercialization of food crops in bogota, Uganda, and the renegotiation of gender. Agender & Society. 1996;**10**(5):608-628

[41] Fischer E, Qaim M. Gender, agricultural commercialization, and collective action in Kenya. Food Security. 2012;**4**(3):441-453. DOI: 10.1007/s12571-012-0199-7

[42] Phommavong S, Ounmany K, Seneduangdeth D, Douangphachanh M, Hathalong K. Women's Empowerment in Coffee Growing in Southern Lao PDR. SAGE Open; Forthcoming

[43] Baird IG. Turning land into capital, turning people into labor: Primative accumulation and the arrival of large-scale economic land concessions in the Lao People's Democratic Republic. New Proposals: Journal of Marxism and Interdisciplinary Inquiry. 2011;**5**(1):10-26

[44] Kenney-Lazar M. Plantation rubber, land grabbing and social-property transformation in southern Laos. Journal of Peasant Studies. 2012;**39**(3-4):1017-1037. DOI: 10.1080/03066150.2012.674942

[45] Toro M. Coffee markets, smallholder credit, and landscape change in the Bolaven Plateau Region, Laos [Master's thesis]. University of Miami; 2012

[46] DoNRE. Land Concession Data. Lao PDR: Department of Natural Resource and Environment; 2013

[47] Lao Statistic Bureau. National Statistics. Vientiane Capital: Lao PDR. Unpublished Report; 2017

[48] MPI. Five Year National Socio-Economic Development Plan VIII (2016-2020). Vientiane Capital: Laos Ministry of Planning and Investment (MPI); 2016

[49] Senedouangdeth D, Ounmany K, Phommavong S, Phouxay K, Hathalong K. Labor employment opportunities in coffee production in southern Lao People's Democratic

Republic. Journal of Asian Rural Studies. 2018;**2**(1):16-36

[50] Phanthouvong S. Efficiency analysis of coffee production in Thateng District, Sekong province, Lao People's Democratic Republic [Master of Economic]. Thailand: Chiengmai University; 2013

Section 6

Coffee By-Products

Chapter 7

Coffee By-Products: Nowadays and Perspectives

Laura Sofía Torres-Valenzuela, Johanna Andrea Serna-Jiménez and Katherine Martínez

Abstract

Coffee is one of the most consumed products around the world; 2.25 billions of coffee cup are consumed everyday in the world. For coffee crop production, different by-products are produced, such as coffee peel, coffee husk, parchment, and spent coffee grounds. These by-products have several problems associated at the final disposition. In this book chapter, we study the main coffee varieties produced in the world, the by-products produced, and its composition and finally assess the potential of supramolecular solvents (SUPRAS) and water as green solvents for high-added-value compound extractions. Bioactive compounds were extracted from fresh and dried coffee peel in an acceptable rate for industrial applications. SUPRAS offer advantages in terms of rapidity (5 min) and simplicity (stirring and centrifugation at room temperature), thus avoiding costly processes based on high pressure and temperature. Extractions carried out using water as solvent is another technique of extraction mixing temperature (above 60°C) and time (4.5 min) obtained a beverage or solution with presence a bioactive compounds how caffeine, chlorogenic acid and polyphenols.

Keywords: agri-food waste, by-products, coffee husks, coffee peel, spent coffee grounds

1. Introduction

Coffee is a tropical perennial plant from the *Coffea* genus of the Rubiaceae family. Although there are more than 103 species recognized nowadays, only 2 are responsible for world trade (arabica and canephora) [1]. The arabica variety constitutes more than 60% of the coffee that is commercialized in the international market and is cataloged by the consumers as the best coffee for its exceptional organoleptic characteristics [2]. This is due to the great variety of chemical compounds, which are responsible for granting the sensory quality and stimuli to the nervous system [3].

Coffee beverage is the result of the preparation of a drink by infusion from roasted and ground beans, with characteristic aroma and flavor, which have made it the second most consumed product in the world [3]. In the case of Colombia, coffee has been cataloged as one of the country's main export products. For the above statement, coffee continues to be an activity of great importance. In this agricultural value chain, the by-products correspond to 80% of the total volume; the coffee industry generates about 2 billion tons of agro-waste, which represent a great pollution hazard [4]. Coffee pulp, husks, silverskin, peel, and spent coffee grounds are common coffee by-products [5].

Generally, coffee is internationally traded as green coffee [6], and it is obtained either by the wet, semi-wet, or dry methods. Typically, wet-processed coffee beans have a higher consumer acceptance than the dry-processed ones [7]. Wet coffee process consists of several steps, namely, de-pulping, fermentation, washing, de-hulling, and drying [8]. Depending on the processing method, either wet or dry, coffee pulp and husk are the first by-products and account for 29 and 12% of the overall coffee cherry [5]. Pulp and husk are rich in carbohydrates (35–85%), soluble fibers (30.8%), mineral (3–11%), proteins (5–11%), and bioactive compounds such as tannins, cyanidins, chlorogenic acid, caffeine and polyphenols [5, 6].

The disposal of agro-waste is a growing issue that can cause phytosanitary problems and cross-contamination in food industries [9]. As a consequence, new strategies to manage or benefit from agro-waste are urgently needed. One of the most promising options is to valorize the bioactive components present in the by-products [10]. In this sense, a growing field of studies highlights the presence of various bioactive compounds in agro-waste with potential applications in functional foods and nutraceutical developments [4, 9, 10]. The recovery of bioactive compounds improves the economic feasibility of the main processes, by producing secondary streams of value-added compounds.

This chapter assesses the usability of SUPRAS and water extraction for the recovery of high-added-value compounds from coffee peel. The method is simple and rapid and could be a sustainable strategy for coffee waste valorization.

2. Origin of coffee

Today, there are countless legends that talk about the origin and discovery of coffee. One of the most accurate ones mentions that coffee originated in the high plateau of Abyssinia and occurred in a wild form known as arabica. It was accidentally discovered by an Ethiopian shepherd named Kaldi. He noticed a strange behavior in his goats when eating fruits and leaves from a certain shrub, so he collected a sample and took it to a monastery [3], where possibly the cherries were mixed in the infusions or thrown into the fire allowing to feel a greater aroma and a better flavor [3].

The Arabs were the first to regularly consume coffee and give a primary role to its cultivation; hence, they are considered the pioneers in the establishment of coffee crops. Subsequently, coffee spreads to Mecca, Medina, and Syria and next to Aden and Cairo, covering the entire Muslim world around 1510 and Turkey in the year 1554 (www.cafedecolombia.com). The introduction of coffee in America was approximately in 1718 starting with the Dutch colony of Suriname, followed by plantations in French Guiana. In 1730, it was the British who introduced coffee into Jamaica and later spread to the rest of the continent [11].

Historically, it has been recognized that coffee was introduced to Colombia via the Venezuelan border by a traveler who came from French Guiana and carried a coffee plant. Thus, the first crops were in the North Santander and, later, in the departments of Antioquia, Tolima, Caldas, Valle del Cauca, Risaralda, Quindío, Cundinamarca, and Nariño, among others. The variety that was initially cultivated in Colombia was the Typica variety. At the end of the 1920s, a second variety was introduced, known as Bourbon, due to its higher yields; however, since the 1980s, the "Colombia" variety has been cultivated, coming from the Caturra variety and the Timor Hybrid, which is resistant to rust [12].

In Colombia, mainly arabica coffee is cultivated, due to this species produce a soft drink and of greater acceptance in the national and international market. The varieties of arabica are low or tall, and have red or yellow fruits. Some varieties of

Variety	Description
Typica	Arabica, pajarito or national Coffee trees are fairly tall Its new leaves or bud are bronzed or reddish. The leaves are elongated Susceptible to rust Greater percentage of large beans than the varieties Caturra and Bourbon Planting density, 2500 trees per hectare
Bourbon	More branches than the Typica variety Lighter green buds than the other leaves Leaves are rounded Produces 30% more than Typica Susceptible to rust Planting density, 2500 trees per hectare
Tabi	Derived from crossing the Timor Hybrid with the Typical and Bourbon varieties Large bean, more than 80% supreme coffee Excellent quality ideal for obtaining specialty coffees Planting density, 3000 trees per hectare Susceptible to rust
Caturra	Lighter green buds than the other leaves Leaves are rounder than Bourbon's Low-to-medium body Produces less than Bourbon and more than Typica Behaves well in the coffee zone Susceptible to rust Planting density, up to 10,000 trees per hectare
Colombia variety	The bud of the plants is bronzed Durable resistance to coffee rust attack Production equal to or greater than Caturra Type of bean and quality of beverage are similar to other varieties of arabica coffee

Table 1.
Colombian coffee varieties.

Arabica species are the Maragogype, Bourbon, Tabi, Typica, Castillo, Caturra, and Colombia, being these last three varieties the ones that are cultivated in greater proportion (see **Table 1**).

Internationally, 80% of the world's production corresponds to the arabica species, which is cultivated mainly in Colombia, Brazil, and in some Asian countries such as India or in Africa such as Kenya and Ethiopia [12]. The remaining 20% corresponds to the species canephora and is generally cultivated in Africa, Brazil, and Indonesia, with differentiating factors such as resistance to rust and a higher caffeine content [13].

The first commercial production of coffee was made in 1835. In this opportunity, 2.560 sack bags were exported from Cucuta. In 1927, the National Federation of Coffee Growers was founded to promote the development of Colombian coffee culture. This organization makes the process of purchasing, storing, and exporting coffee, as well as accompanying and advising coffee growers from different regions of the country. In this way, coffee cultivation was consolidated as one of the country's main agricultural activities [3]. Today, the sector continues to be an important articulating axis in the country's rural development and providing economic stability despite the coffee crises represented by high production costs and low harvest levels. To date, the National Federation of Coffee Growers has reported a participation of 560,000 farms dedicated to coffee cultivation, which translates to 948,000 hectares of which 27% are harvested with the variety Colombia. The rest corresponds mainly to the varieties Typica, Caturra, and Bourbon [12], with 66% of the cultivated area in the country, being cataloged as the product with greater participation among the other registered crops [14], providing approximately 785 thousand rural jobs directly and 1.5 million indirectly (www.federaciondecafeteros.org).

3. Characteristics of coffee

Coffees are berries obtained from a perennial and topical plant (cafeto) [3]. The coffee beans are morphologically very variable and have different shapes, colors, and sizes. Internally, seeds are found (usually two per fruit), which are processed and used to prepare infusions [3].

Plants have a cleft in the central part of the seed. Depending on the species, it is possible to find small shrubs or trees larger than 10 m. The leaves are simple, opposite, and with stipules and present variability both in size and texture. The plant has white and tubular flowers, which are complete, i.e., all organs are in the same flower [2]. The root is a vital organ for the coffee plant, because through it, the plant takes the water and nutrients for its growth and also is an anchor to the soil [3]. The coffee plant has a main root that can reach a depth of up to 50 cm, from which other thick roots are available to support the thinner ones in charge of absorbing nutrients. The stem forms the skeleton of the coffee tree along with the branches with leaves, flowers, and finally, the fruits (www.cafedecolombia.com).

Because of the union of the grain of pollen with the ovule, the fruit and seeds are formed. The coffee fruit is a cherry that is divided into three layers: epicarp or skin, which is the outermost layer; mesocarp or pulp, which forms a sweet and aromatic pulp of mucilaginous nature, protected by a yellow cellulose layer called parchment or endocarp; and finally a silvery layer, which covers the two oval-shaped grains called endosperm [11].

3.1 Chemical composition of coffee

Coffee has a number of chemical components, mainly water and dry matter, such as minerals, organic substances (carbohydrates, lipids, proteins), alkaloids (caffeine and trigonelline), carboxylic and phenolic acids, and volatile compounds responsible for the aroma. All together result in a great diversity and complexity of structures; however, these may have modifications in any of their stages, either from the crop or the mill [15].

The chemical composition varies depending on the species [16]. In the case of *Coffea arabica*, it has a higher lipid and sucrose content than *Coffea canephora*. The robusta differs by its higher content of polysaccharides, caffeine, chlorogenic acids, and ashes. **Table 2** shows the most representative chemical components in arabica and robusta species.

Additionally, within the varieties cultivated in Colombia are differences (see **Table 3**), due to the intrinsic factors, soil fertilization, atmospheric conditions, sowing density, and planting age, among others [15].

Water: The water content of the bean is one of the most relevant factors in all coffee processes, from germination to roasting. In the fresh fruit, the water content is between 70 and 80% [25]. After the dry process, the water content is reduced up to 10–12% to improve the stability and avoid microbial proliferation, prolonging its shelf life [16].

Carbohydrates: Among the main polysaccharides in coffee are mannan or galactomannan (polymer of mannose and galactose), constituting 50% of the polysaccharides, 30% of arabinogalactan (polymer of galactose and arabinose), 15% of cellulose (polymer of glucose), and 5% of peptic substances [16]. The beans in an optimum

Chemical component*	Puerta [16]		Echeverry et al. [17]		Komes [18]	
	Arabica	Robusta	Arabica	Robusta	Arabica	Robusta
Polysaccharides	50.8	56.40	38	41.5	N.D	N.D
Sucrose	8.00	4.00	N.D	N.D	N.D	N.D
Reducing sugars	0.10	0.40	N.D	N.D	N.D	N.D
Protein	9.80	9.50	10	1o	N.D	N.D
Amino acids	0.50	0.80	N.D	N.D	N.D	N.D
Caffeine	1.20	2.20	1.3	1.4	0.76–1.82	1.51–3.33
Trigonelline	1.00	0.70	1	0.7	0.88–2.76	0.75–3.42
Lipids	16.20	10.00	17	11	N.D	N.D
Aliphatic acids	1.10	1.20	2.4	2.5	N.D	N.D
Chlorogenic acids	6.90	10.40	2.7	3.1	4–8.4	7–14.4
Minerals	4.20	4.40	4.5	4.7	N.D	N.D
Aromatic compounds	Traces	Traces	0.1	0.1	Traces	Traces
Melanoidins	N.D	N.D	23	23	25	25

**Expressed in percentage, on a dry basis.*
N.D. Non-determined.

Table 2.
Chemical composition of the coffee beans of the arabica and robusta species.

Coffee variety	Fiber (%)	Lipids (%)	Proteins (%)	Caffeine (%)	Chlorogenic acids (%)	Ash (%)
Bourbon	21.75	15.27	13.90	1.15	7.37	3.78
Caturra	18.85	13.98	14.79	1.13	6.97	3.39
Colombia yellow fruit	18.45	13.07	14.45	1.16	7.55	3.49
Colombia red fruit	16.69	14.27	13.92	1.19	7.42	3.52
Typica	18.71	13.99	14.59	1.20	6.66	3.43
Robusta	15.53	11.42	15.66	2.10	8.08	3.96

Source: Puerta [16].

Table 3.
Chemical composition of coffee bean in the different varieties sown in Colombia.

ripening stage have a higher sucrose content than defective and immature beans. In the arabica species, the sucrose content ranged between 6 and 9%, while robusta contains 3–7% of sucrose [16]. Monosaccharides and some disaccharides such as lactose and maltose may oxidize to form alcohols and acids in the fermentation process or may react with the amino acids in roasting to form melanoidins, which are responsible for the coloration (enzymatic browning) of the roasted coffee [16].

Lipids: Triglycerides, linoleic, and palmitic acid are mainly presents (~ 75% of coffee lipids). The unsaponifiable matter constitutes 20 to 25% of the lipids of coffee. Sterols are 2.2% of coffee lipids and contain β-sitosterol, stigmasterol, campesterol, and Δ5 avenasterol. Cholesterol constitutes 0.11% of the dry weight of coffee beans (0.044% in robusta coffee) [16].

Nitrogen compounds: Nitrogen constitutes between 1.30 and 3.23% of the dry weight of the green coffee beans, after the roasted decreased up to 1.51 and 2.14% [16].

Alkaloids: Alkaloids are the substances responsible for giving the bitter taste of coffee, the most representative are caffeine, trigonelline, paraxanthine, theobromine, and theophylline [16]. Caffeine is a methylxanthine, which have attributed health benefits, such as improve the central nervous, cardiovascular, respiratory, renal, and muscular system [19]. For the above statement, caffeine is important in the pharmaceutical industry. Also it has important bioactive properties, so it may be cataloged as a functional ingredient, which can be used in different food matrices [20].

Chlorogenic acids: They are a series of phenolic esters derived from the union of an ester between caffeic acid and quinic acid [3]. The chlorogenic acid content in green coffee is 7% and reaches 4% after roasting [3]. A volume of 200 mL of roasted and ground coffee could provide between 70 and 350 mg of chlorogenic acid [16]. Coffee beans contain more than 40 chlorogenic acids, especially esters of quinic acid such as CQA, di-CQA, and FQA [16]. This compound has a significant antioxidant capacity and also a stimulant, expectorant, diuretic, choleretic, and antihepatotoxic effects [21].

3.2 By-products of coffee processing

The coffee bean is picked after reach the commercial ripening stage; it next must be quickly transformed into dry parchment coffee, to avoid accelerated fermentation because the entire bean includes high water and sugar content [22]. For these purposes the external layer is removed from the coffee bean and only 5% of the biomass is used to produce a coffee crop, the rest remains in a residual form as leaves, branches, green fruits, pulp, mucilage, parchment, and silverskin, among others [22]. There are two primary methods for processing coffee, to obtain green coffee (traded coffee beans): wet and dry. In the dry process, no layers are removed, and coffee cherries are laid out in the sun to dry. In the wet process, the fruit covering the layer is removed before they are dried. Approximately 40% of all coffee around the world is wet processed [23], because it is considered to produce superior tasting offers [8, 24]. In the wet process, it has been estimated that 40–45 L of wastewater are produced per kilogram of coffee [25].

In Colombia, the wet process has been implemented for decades, which generates a contamination of 115 g of COD per kilogram of cherry coffee [22]. To overcome this problem, new methods were developed; one of these is the Belcosub technology, in which the fruit is de-pulped. The external layer is transported without water, and the organic residues are reused; however, these do not generate a significant value. This system avoids up to 74% of the contamination of water resources, since less than 5 l/kg of dry parchment are used [22]. The most recent technology suggested by the National Federation of Coffee Growers is the ecological mill without dumping (Ecomil) that reduces the amount of water to 0.5 l of water per kilogram of dry parchment, implementing tanks generally in stainless steel that do not need water for coffee emptying. In addition, the water resulting from this process goes directly to purifying tanks with microorganisms and a series of filters that allow the water that falls to water sources to be clean and do not generate any pollution [22].

As mentioned above, a large amount of coffee bean components are removed. It is important to highlight that approximately 43.58% of the weight of the dried fruit are these by-products [22]. The valorization of these by-products through the recovery of bioactives has increasingly become of interest for food, pharmaceutical, and cosmetic industries [26–30].

A promising option to recover these bioactive compounds is the coffee pulp, which involves the epicarp and part of the mesocarp of the fruit. This by-product

By-product	Process	Final use	References
Fresh pulp	Dried	Biocomponents	[40]
Fresh pulp	Chopped	Animal feed	[41, 42]
	Silage		[43]
Fresh and dried pulp	Washing, drying, blending, and freezing extracts with hot water	Biocomponents extraction, antioxidant activity, and bacterial inhibitory activity	[44]
Husk	Supercritical carbon dioxide	Extraction of caffeine	[45]
Fresh and dried pulp	Mixtures of solvents with ethanol and milli-Q water in different mixtures and sonication—maceration	Antioxidant activity, cytotoxicity on human epithelial gastric cells, and biocomponents profile	[46]
Silverskin	Subcritical water (25–270°C)	Antioxidant activity	[47]
	Batch culture fermentation bifidobacteria	Prebiotic potential dietary fiber	[48]
Parchment	Water extraction, concentration, and dried	Inhibitory effect of a on hyaluronidase	[49]
	Ethanol extraction mixed with temperature	Biocomponents extraction, antioxidant activity	[50]
Fresh pulp	Fermentation	Ethanol	[51, 52]
Fresh pulp and mucilage			[53]
Parchment	Hydrolysis and fermentation		[54]
Fresh pulp	Pretreatment with *Mycotypha* sp. and biomethanization	Biogas	[55]
	Pretreatment with *Streptomyces* sp. and biomethanization		[56]
Parchment	Hydrolysis and biomethanization		[57]
Pulp and rejected grain			[58]
Fresh and dried pulp	Co-digestion biomethanization		[59–63]
Parchment	Microwave and traditional heating	Pyrolysis	[64]
Husks	Combustion furnace as well as in a fluidized bed combustion chamber at pilot scale		[65]
	Vermicomposting with *Eudrilus eugeniae* and *Trichoderma viride*	Compost	[66]
Parchment	Composted parchment with a mixture of organic amendments, pulp with bovine manure, and the legume *Millettia ferruginea*		[67]
Pulp	Compost with *Trichoderma* sp., *Streptomyces* sp. *Azotobacter* sp., and *Bacillus* sp.		[68]
Husk	Solid-state fermentation with *Rhizopus, Phanerochaete*, and *Aspergillus* sp.	Evaluate the reduction of caffeine and tannins	[69]

By-product	Process	Final use	References
Dried coffee leaves, coffee cherry husk, coffee parchment skin, silverskin, and spent coffee ground	Solid-state fermentation	Produced the fungus *Pleurotus florida*	[70]
Pulp		Production of an extract rich in chlorogenic acid	[71]
Coffee husk and pulp dried	Hydrolysis and sterilization; fermentation using *Rhodotorula mucilaginosa*	Carotenoids production	[72]

Table 4.
Alternatives for added value of coffee pulp.

contains significant amounts of caffeine and another component [31]. The reported chemical composition (expressed in dry mass) includes polyphenols (1.5–2.9%), total sugars (4.1%), protein (4–13.3%), lignin (17.5–19.3%), lipids (1.7–2.5%), cellulose (18–63%), total fiber (18–60.5%), ash (6–10%), tannins (1.8–9%), carbohydrates (44–89%), reducing sugars (12.4%), nonreducing sugars (2%), caffeine (1.2–1.5%), and chlorogenic acid (1.6%) [5, 31–33].

3.3 Nowadays uses of coffee by-products

Coffee consumption has increased significantly, which generates an increase in waste amount [32]. These wastes have pollution problems. Discharges of wastewater from industrial activities have become a global issue of concern [34]. Different alternatives to both mitigate the negative effects in the discharges of coffee by-products and generate added-value alternatives have been evaluated. For this purpose, different studies have been carried out to evaluate alternative uses and reduce the toxic effect on the environment [35].

According to CONAMA Resolution No. 430 from 05/13/2011, the concentration of phenols should be lower than 0.5 mg L^{-1} [36], due to when phenolic compounds are discharged into the environment will lead to the degradation process of organic materials difficult to degrade [37]. Coffee by-products are the polyphenols and also carbohydrates, proteins, and pectins, making them potential sources of agro-industrialization for various industries, as well as renewable economic resources that can be given a high added value [32]. **Table 4** presented alternatives for coffee by-product reuse.

Among the alternatives previously proposed for the food and nonfood industry, emphasis will be placed on the production of extraction of caffeine both from the beans and from the different by-products obtained.

Considering that caffeine is an alkaloid that possesses antioxidant capacity and increases energy availability, cognitive performance, and neuromuscular coordination, among others [38], one of the alternative uses of coffee and its by-products has focused on the extraction of this compound of functional interest, both for the food and nonfood sectors. However, in recent years, new extraction techniques have been sought in order to reduce the generation of waste, with a lower consumption of chemical reagents and therefore improve the efficiency of the process, reducing process times [39].

4. SUPRAS extraction

Two amphiphiles (decanoic acid and hexanol) and two dispersion solvents (THF:water and ethanol:water) were selected for the study to generate a variety of

SUPRAS based on previous promising results [73–75]. Hexanol has the advantage over decanoic acid for the possibility of removal by evaporation if further steps are required after the extraction of bioactives, while decanoic acid is a more biocompatible and renewable option. Thus, this offers different strategies for amphiphile recovery and reutilization in industrial purposes, being usually easier the operation with liquid phases. Ethanol was tested together with THF, the first considered biocompatible and authorized for use in food and the second easily removable by evaporation due to its high vapor pressure (143 mm Hg at 20°C) and relatively low boiling point (66°C). The coacervating agent (external stimuli driving the self-assembly synthesis of SUPRAS) is water in both cases, as a poor solvent for the amphiphiles promoting the aggregation as described before [76]. The type of amphiphile and of the dispersion solvent and the composition of the ternary mixture in the bulk solution (amphiphile, organic solvent, water) give rise to SUPRAS with different final composition and microstructure and volumes which can influence the extraction efficiency [73]. Thus, SUPRAS binding interactions and restricted access properties (conferred by the size of the aggregates) can be tuned depending on the amphiphile functional groups (OH, COOH) providing hydrogen bonds for extraction and dipole–dipole interactions, the alkyl chain length (C6, C10) giving dispersion interactions and the dispersion medium composition (providing hydrogen bonds, dipole–dipole interactions, and dispersion forces under different ratios of ethanol:water or THF:water mixtures).

Figures 1 and **2** show the caffeine and chlorogenic acid extraction with SUPRAS. As expected, recoveries were influenced by the amphiphile nature and bulk solution composition of the ternary mixture (amphiphile, water, organic solvent). Under all the tested SUPRAS, the caffeine extraction efficiency was in the range 31–68% ± 2.8%, while the chlorogenic acid extraction efficiency was between 0 and 26 ± 2.5%.

The highest caffeine extraction was obtained with hexanol as amphiphile and ethanol:water as dispersion solvent for the design of the SUPRAS (maximum at

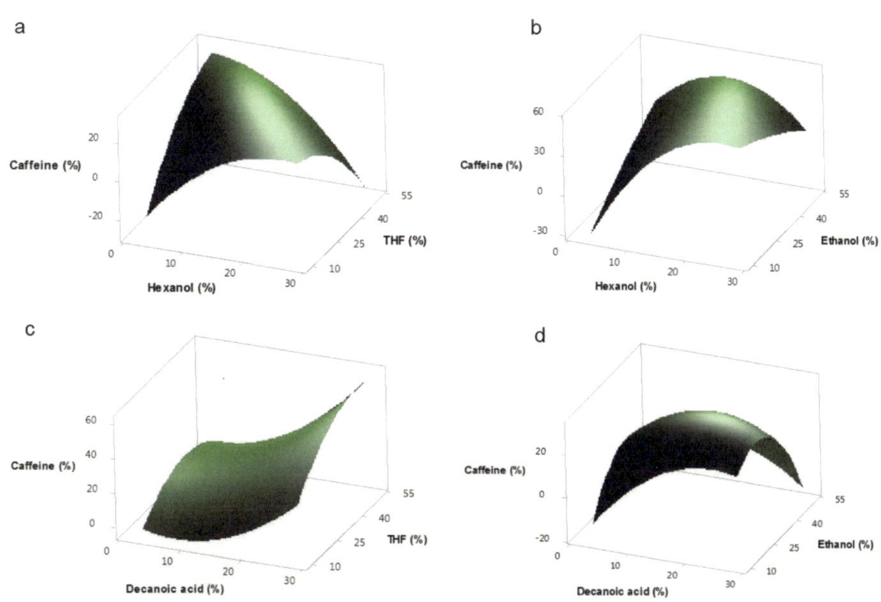

Figure 1.
Surface response plots for caffeine extraction from coffee peel with (a) hexanol, THF; (b) hexanol, ethanol; (c) decanoic acid, THF; and (d) decanoic acid, ethanol.

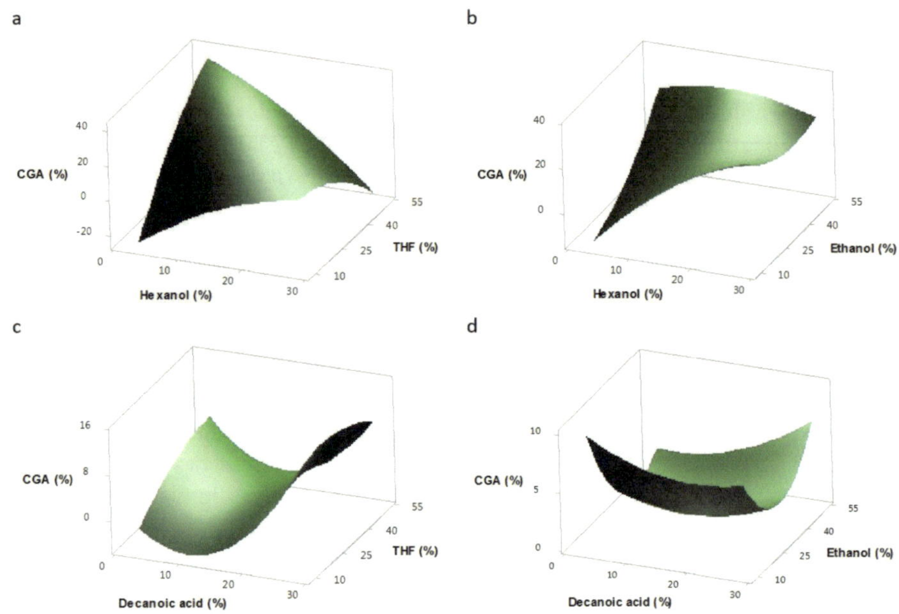

Figure 2.
Surface response for chlorogenic acid extraction from coffee peel with (a) hexanol, THF; (b) hexanol, ethanol; (c) decanoic acid, THF; and (d) decanoic acid, ethanol.

69 ± 0.9% with 7% of hexanol and 15% of ethanol). The range obtained for the other SUPRAS was 45–56 ± 1.1%, 31–56 ± 2%, and 39–65 ± 7.5% with hexanol-THF, decanoic acid-THF, and decanoic acid-ethanol, respectively.

SUPRAS based on ethanol:water were indeed more suitable than those based on THF:water to extract caffeine with both amphiphiles. A possible explanation is that ethanol as protic solvent can extract caffeine more efficiently than THF (aprotic solvent), acting as hydrogen bond donor for caffeine, which contains hydrogen bond acceptors' groups only. Additionally, the dielectric constant of ethanol is higher than that of THF (24 and 7.5, respectively). This parameter is a relative measure of the chemical polarity and could enhance the extraction of the polar bioactives by dipole–dipole interactions. With respect to the amphiphile, hexanol was the best choice, and the highest efficiency rates (69 ± 0,9%) were obtained in SUPRASs formed with this organic alcohol. The higher polarity of hexanol over decanoic acid could be the reason for the higher extraction efficiency of the polar bioactive compounds. Furthermore, the smaller size of hexanol aggregates, due to its shorter alkyl chain length, could generate SUPRAS with greater surface area, and, consequently, it will provide more available binding interactions for the bioactive components.

Chlorogenic acid extraction rates were lower than caffeine rates, and no clear correlation was found with the SUPRAS synthetic conditions. Its higher polarity could lead to losses in the equilibrium competing phase (i.e., calculated log P −0.4 and −0.1 for chlorogenic acid and caffeine, respectively). Furthermore, chlorogenic acid is a bigger molecule than caffeine; its molar mass is 354.31 g/mol, and its topological polar surface area is 165 A2, while for caffeine, values of 194.19 g/mol and 58.4 A2 are calculated. The higher contact polar area of caffeine could enhance the recoveries too. Caffeine is the most routinely ingested bioactive substance. Its consumption possesses health benefits, including lower risks of Parkinson's and Alzheimer's disease, a favorable effect on liver function, energy expenditure, and a decreased risk of developing certain cancers (endometrial, prostatic, colorectal, liver) [77]; it can stimulate fat oxidation, thermogenesis, and energy expenditure subsequently, which reduces body weight [78]. Caffeine is consumed daily, in the United States 89% of the population 19 years of age or older consumes some form of caffeine daily, but the primary source of caffeine is coffee (64%) [79].

Chlorogenic acid is the major polyphenol in edible plants with many health-promoting properties [80]. It has a strong antioxidant activity, anti-lipid peroxidation, anticancer effects [81], anti-inflammatory activity, inhibition of α-amylase, and α-glucosidase linked to type 2 diabetes and anti-obesity properties [82]; it also has antimicrobial properties [83]. Due to the beneficial effects of this bioactive component, it has been used for the preparation of functional materials in food and pharmaceutical areas [80].

5. Water extraction

The processing of every 60,000 tons of dried coffee beans produces approximately 218,400 tons of fresh pulp and mucilage or mesocarp [84]. Generally, the pulp is removed with mechanical movements generated by pulping and constitutes about 29–43% (w/w) of the fruit [6, 85]; the pulp a potential use has been identified by the compounds present such as anthocyanins, caffeine, and phenolic compounds with which an important added value can be generated [46, 86, 87]. In this study dried pulp was employed for the biocomponents extractions, using hot water as solvent, the dried pulp of arabica variety was selected with 10–12% of humidity, the response surface methodology was used to determine the effect of solvent temperature (water) (60–90°C) and extraction time (1–8 min) on the functional characteristics of the infusions obtained.

For the preparation of the infusions, dried pulp was taken and placed in infusers. Each sample was deposited in a beaker with 250 mL of the solvent (water) at a different time and temperature conditions. The samples were quantified polyphenol content by the Folin–Ciocalteu method reported by [44, 46]; the quantification of caffeine and chlorogenic acid was done by high-performance liquid chromatography (HPLC).

The chromatographic separation was performed in a Shimadzu Prominence with a UV detector and quaternary pump system (Shimadzu, Japan); the samples were filtrated in a cellulose filter of 25 μm, and the filtrated sample (20 μL) was conducted using a C8 Restek column (Restek Corporation, USA). The mobile phase consisted of 0.1% acetic acid and 30% methanol in water v/v; the injection volume was 20 μL. The mobile phase flow rate was 0.5 mL/min (35°C). The reference standards were used for identification, and calibration curves were obtained for quantification chlorogenic acid and caffeine.

The peak of caffeine was observed at the elution time of 11.59 min. The caffeine extracted from 3.3 g of coffee pulp ranged between 21–51 mg/L and did not depend on the extraction temperature from 65 to 90°C, the time has an effect in time upper 4.5 min [47], and the values of caffeine were higher (**Figure 3a**). The chlorogenic acid had a similar behavior of caffeine (**Figure 3b**) with range values 5–9 mg/L; this indicates that those substances are stable during extraction and heat treatment and storage of the beverage [84].

In the extraction process, this type of biocomponents is the solvent, since the type of compound to be extracted depends on the type of solvent used for the capacity they possess which is directly related to their polarity. Extractions using water improved the extraction of phenolic compounds, caffeine and chlorogenic acid due to it polarity [84, 88].

Therefore, coffee pulp can be a raw material with a high content of compounds, and its consumption (e.g., in infusions or extracts) can help prevent degenerative diseases, taking into account that a relationship has been established between

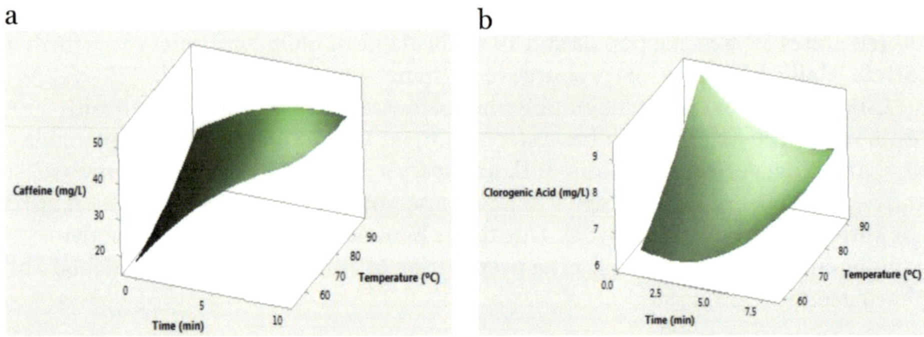

Figure 3.
Surface response for (a) caffeine and (b) chlorogenic acid using water as solvent.

consumption of biocomponents such as polyphenols, caffeine, and chlorogenic acid and the reduction of risks of chronic diseases, including obesity and diabetes [5, 6, 32, 89]. The coffee pulp has potential for use in the food, pharmaceutical, and cosmetic industry, becoming an alternative for products to generate added value and reduce negative effects on the environment and improve the profitability of producers within of circular economy and biorefineries.

6. Conclusion

This study shows the ability of SUPRAS, nanostructured solvents made up of assembled amphiphile aggregates and water, for valorization of coffee waste. The results proved that these solvents offer excellent extraction capacity of high-added-value compounds with interest for the food, pharmaceutical, and cosmetic industry.

Author details

Laura Sofía Torres-Valenzuela*, Johanna Andrea Serna-Jiménez and Katherine Martínez
University of Cordoba, Cordoba, Spain

*Address all correspondence to: torresvallaura@miugca.edu.co

References

[1] Naranjo M, Vélez LT, Rojano BA. Actividad antioxidante de café colombiano de diferentes calidades. Revista Cubana de Plantas Medicinales. 2011;**16**(2):164-173

[2] Suarez Agudelo JM. Aprovechamiento de los residuos sólidos provenientes del beneficio de café, en el municipio de Betania Antioquia: Usos y aplicaciones. Caldas: Corporación Universitaria Lasallista; 2012

[3] Gotteland M, de Pablo VS. Algunas verdades sobre el café. Revista Chilena de Nutricion. 2007;**34**(2):105-115

[4] Bandara N, Chalamaiah M. Bioactives from agricultural processing by-products. In: Reference Module in Food Science. Elsevier; 2018. [citado 24 de julio de 2018]. Disponible en: http://www.sciencedirect.com/science/article/pii/B9780081005965224086

[5] Janissen B, Huynh T. Chemical composition and value-adding applications of coffee industry by-products: A review. Resources, Conservation and Recycling. 2018;**128**:110-117

[6] Esquivel P, Jiménez VM. Functional properties of coffee and coffee by-products. Food Research International. 2012;**46**(2):488-495

[7] Worku M, de Meulenaer B, Duchateau L, Boeckx P. Effect of altitude on biochemical composition and quality of green arabica coffee beans can be affected by shade and postharvest processing method. Food Research International. 2018;**105**:278-285

[8] Dadi D, Mengistie E, Terefe G, Getahun T, Haddis A, Birke W, et al. Assessment of the effluent quality of wet coffee processing wastewater and its influence on downstream water quality. Ecohydrology and Hydrobiology. 2018;**18**(2):201-211

[9] Lai WT, Khong NMH, Lim SS, Hee YY, Sim BI, Lau KY, et al. A review: Modified agricultural by-products for the development and fortification of food products and nutraceuticals. Trends in Food Science and Technology. 2017;**59**:148-160

[10] Bhatnagar A, Sillanpää M, Witek-Krowiak A. Agricultural waste peels as versatile biomass for water purification—A review. Chemical Engineering Journal. 2015;**270**:244-271

[11] Sánchez DA, Anzola C. Caracterización química de la película plateada del café (*Coffea arábica*) en variedades colombia y caturra. Revista Colombiana de Química. 2012;**41**(2):211-226

[12] Puertas Quintero GI. Calidad en taza de las variedades de Coffea arabica L. cultivadas en Colombia. Cenicafé. 1998;**49**(4):265-278

[13] Sarasty Zambrano DJ. Alternativas de tratamiento del mucilago residual producto del beneficiadero de café [postgrade thesis]. Colombia: Universidad Industrial de Santander; 2012 [citado 9 de agosto de 2019]. Disponible en: https://studylib.es/doc/6437128/alternativas-de-tratamiento-del-mucilago-residual-product...

[14] Pérez Toro JA. Economía cafetera y desarrollo económico en Colombia. Bogotá, Colombia: Universidad Jorge Tadeo Lozano; 2013. 592 p

[15] Montilla Pérez J, Arcila-Pulgarín J, Aristizábal-Loaiza M, Montoya-Restrepo EC, Puertas Quintero GI, Oliveros-Tascón CE, et al. Caracterización de algunas propiedades físicas y factores de conversión del café

durante el proceso de beneficio húmedo tradicional. Cenicafé. 2008;**59**:2

[16] Puertas Quintero GI. Composición química de una taza de café. Av Tecnológicos Cenicafé. 2011;**414**:1-12

[17] Echeverry D, Buitrago L, Montes F, Mejía I, González MP. Café para cardiólogos. Revista Colombiana de Cardiología. 2005;**11**(8):357-365

[18] Komes D, Bušić A. Chapter 3—Antioxidants in coffee. In: Preedy V, editor. Processing and Impact on Antioxidants in Beverages [Internet]. San Diego: Academic Press; 2014. pp. 25-32. Disponible en: http://www.sciencedirect.com/science/article/pii/B9780124047389000039

[19] Institute of Medicine (US) Committee on Military Nutrition Research. Caffeine for the Sustainment of Mental Task Performance: Formulations for Military Operations [Internet]. Washington, DC: National Academies Press (US); 2001 [citado 12 de agosto de 2019]. Disponible en: http://www.ncbi.nlm.nih.gov/books/NBK223802/

[20] Chu Y-F, Chen Y, Brown PH, Lyle BJ, Black RM, Cheng IH, et al. Bioactivities of crude caffeine: Antioxidant activity, cyclooxygenase-2 inhibition, and enhanced glucose uptake. Food Chemistry. 2012;**131**(2):564-568

[21] Fonnegra GR, Jiménez RSL. Plantas medicinales aprobadas en Colombia. Medellín, Colombia: Universidad de Antioquia; 2007. 394 p

[22] Rodríguez Valencia N, Sanz Uribe JR, Oliveros-Tascón CE, Ramírez Gómez CA. Beneficio del café en Colombia. Chinchiná, Colombia: Cenicafé; 2015. 35 p

[23] Garde WK, Buchberger SG, Wendell D, Kupferle MJ. Application of *Moringa oleifera* seed extract to treat coffee fermentation wastewater. Journal of Hazardous Materials. 2017;**329**:102-109

[24] Aguiar LL, Andrade-Vieira LF, de Oliveira David JA. Evaluation of the toxic potential of coffee wastewater on seeds, roots and meristematic cells of *Lactuca sativa* L. Ecotoxicology and Environmental Safety. 2016;**133**:366-372

[25] Zayas Péerez T, Geissler G, Hernandez F. Chemical oxygen demand reduction in coffee wastewater through chemical flocculation and advanced oxidation processes. Journal of Environmental Sciences. 2007;**19**(3):300-305

[26] Campos-Vega R, Loarca-Piña G, Vergara-Castañeda HA, Oomah BD. Spent coffee grounds: A review on current research and future prospects. Trends in Food Science and Technology. 2015;**45**(1):24-36

[27] Hall S, Desbrow B, Anoopkumar-Dukie S, Davey AK, Arora D, McDermott C, et al. A review of the bioactivity of coffee, caffeine and key coffee constituents on inflammatory responses linked to depression. Food Research International. 2015;**76**(3):626-636

[28] Kovalcik A, Obruca S, Marova I. Valorization of spent coffee grounds: A review. Food and Bioproducts Processing. 2018;**110**:104-119

[29] McNutt J, He QS. Spent coffee grounds: A review on current utilization. Journal of Industrial and Engineering Chemistry. 2019;**71**:78-88

[30] Torres-Valenzuela LS, Ballesteros-Gómez A, Sanin A, Rubio S. Valorization of spent coffee grounds by supramolecular solvent extraction. Separation and Purification Technology. 2019;**228**:115759

[31] Armas Flores EA, Cornejo Mazariego NC, Murcia Zamora KM. Propuesta para el aprovechamiento de los subproductos del beneficiado del café como una alternativa para la diversificación de la actividad cafetera y aporte de valor a la cadena productiva. Buenos Aires, Argentina: Universidad del Salvador; 2008

[32] Murthy PS, Madhava Naidu M. Sustainable management of coffee industry by-products and value addition—A review. Resources, Conservation and Recycling. 2012;**66**:45-58

[33] Preedy VR. Coffee in Health and Disease Prevention [Internet]. Elsevier; 2015 [citado 9 de julio de 2019]. 1080 p. Disponible en: https://linkinghub.elsevier.com/retrieve/pii/C20120069591

[34] Anastopoulos I, Karamesouti M, Mitropoulos AC, Kyzas GZ. A review for coffee adsorbents. Journal of Molecular Liquids. 2017;**229**:555-565

[35] Noriega Salazar A, Silva Acuña R, García de Salcedo M. Composición química de la pulpa de café a diferentes tiempos de ensilaje para su uso potencial en la alimentación animal. Zootecnia Tropical. 2009;**27**:135-141

[36] Chagas PMB, Torres JA, Silva MC, Corrêa AD. Immobilized soybean hull peroxidase for the oxidation of phenolic compounds in coffee processing wastewater. International Journal of Biological Macromolecules. 2015;**81**:568-575

[37] Novita E. Biodegradability simulation of coffee wastewater using instant coffee. Agriculture and Agricultural Science Procedia. 2016;**9**:217-229

[38] Guglielmetti A, D'Ignoti V, Ghirardello D, Belviso S, Zeppa G. Optimisation of ultrasound and microwave-assisted extraction of caffeoylquinic acids and caffeine from coffee silverskin using response surface methodology. Italian Journal of Food Science. 2017;**29**(3). Disponible en: https://www.chiriottieditori.it/ojs/index.php/ijfs/article/view/727

[39] De Marco I, Riemma S, Iannone R. Life cycle assessment of supercritical CO_2 extraction of caffeine from coffee beans. Journal of Supercritical Fluids. 2018;**133**:393-400

[40] Torres-Valenzuela LS, Martínez KG, Serna-Jimenez JA, Hernández MC. Secado de pulpa de café: Condiciones de proceso, modelación matemática y efecto sobre propiedades fisicoquímicas. Información Tecnológica. 2019;**30**(2):189-200

[41] Pedraza-Beltrán P, Estrada-Flores JG, Martínez-Campos AR, Estrada-López I, Rayas-Amor AA, Yong-Angel G, et al. On-farm evaluation of the effect of coffee pulp supplementation on milk yield and dry matter intake of dairy cows grazing tropical grasses in central Mexico. Tropical Animal Health and Production. 2012;**44**(2):329-336

[42] Didanna HL. A critical review on feed value of coffee waste for livestock feeding. World Journal of Biology and Biological Sciences. 2014;**2**(5):72-086

[43] Rathinavelu R, Graziosi G. Potential Alternative Use of Coffee Wastes and by-Products. 2005:1-4

[44] Duangjai A, Suphrom N, Wungrath J, Ontawong A, Nuengchamnong N, Yosboonruang A. Comparison of antioxidant, antimicrobial activities and chemical profiles of three coffee (*Coffea arabica* L.) pulp aqueous extracts. Integrative Medicine Research. 2016;**5**(4):324-331

[45] Tello J, Viguera M, Calvo L. Extraction of caffeine from Robusta coffee (*Coffea canephora* var. Robusta) husks using supercritical carbon

dioxide. Journal of Supercritical Fluids. 2011;**59**:53-60

[46] Magoni C, Bruni I, Guzzetti L, Dell'Agli M, Sangiovanni E, Piazza S, et al. Valorizing coffee pulp by-products as anti-inflammatory ingredient of food supplements acting on IL-8 release. Food Research International. 2018;**112**:129-135

[47] Narita Y, Inouye K. High antioxidant activity of coffee silverskin extracts obtained by the treatment of coffee silverskin with subcritical water. Food Chemistry. 2012;**135**(3):943-949

[48] Borrelli RC, Esposito F, Napolitano A, Ritieni A, Vicenzo F. Characterization of a new potential functional ingredient: Coffee silverskin. Journal of Agricultural and Food Chemistry. 2004;**52**(5):1338-1343

[49] Furusawa M, Narita Y, Iwai K, Fukunaga T, Nakagiri O. Inhibitory effect of a hot water extract of coffee "silverskin" on hyaluronidase. Bioscience, Biotechnology, and Biochemistry. 2011;**75**(6):1205-1207

[50] Mirón-MéridaVA,Yáñez-FernándezJ, Montañez-Barragán B, Barragán Huerta BE. Valorization of coffee parchment waste (*Coffea arabica*) as a source of caffeine and phenolic compounds in antifungal gellan gum films. LWT-Food Science and Technology. 2019;**101**:167-174

[51] Gouvea BM, Torres C, Franca AS, Oliveira LS, Oliveira ES. Feasibility of ethanol production from coffee husks. Biotechnology Letters. 2009;**31**(9):1315-1319

[52] Bonilla VA. Reaproveitamento de resíduos da indústria do café como matéria prima para a produção de etanol. Universidad Federal de Lavras; 2014

[53] Harsono SS, Salahuddin, Fauzi M, Purwono GS, Soemarno D, Kissinger. Second generation bioethanol from Arabica coffee waste processing at smallholder plantation in Ijen plateau region of East Java. Procedia Chemistry. 2015;**14**:408-413

[54] Mussatto SI, Machado EMS, Carneiro LM, Teixeira JA. Sugars metabolism and ethanol production by different yeast strains from coffee industry wastes hydrolysates. Applied Energy. 2012;**92**:763-768

[55] Jayachandra T, Venugopal C, Anu Appaiah KA. Utilization of phytotoxic agro waste—Coffee cherry husk through pretreatment by the ascomycetes fungi Mycotypha for biomethanation. Energy for Sustainable Development. 2011;**15**(1):104-108

[56] Orozco AL, Pérez MI, Guevara O, Rodríguez J, Hernández M, González-Vila FJ, et al. Biotechnological enhancement of coffee pulp residues by solid-state fermentation with *Streptomyces*. Py–GC/MS analysis. Journal of Analytical and Applied Pyrolysis. 2008;**81**(2):247-252

[57] Kivaisi A. Pretreatment of robusta coffee hulls and co-digestion with cow-dung for enhanced anaerobic digestion. Tanzania Journal of Science. 2002;**28**(2):1-10

[58] Battista F, Fino D, Mancini G. Optimization of biogas production from coffee production waste. Bioresource Technology. 2016;**200**:884-890

[59] Corro G, Paniagua L, Pal U, Bañuelos F, Rosas M. Generation of biogas from coffee-pulp and cow-dung co-digestion: Infrared studies of postcombustion emissions. Energy Conversion and Management. 2013;**74**:471-481

[60] Houbron E, Larrinaga A, Rustrian E. Liquefaction and methanization of solid and liquid coffee

wastes by two phase anaerobic digestion process. Water Science and Technology. 2003;**48**(6):255-262

[61] Chala B, Oechsner H, Latif S, Müller J. Biogas potential of coffee processing waste in Ethiopia. Sustainable Switzerland. 2018;**10**(8):1-14

[62] Neves L, Oliveira R, Alves MM. Anaerobic co-digestion of coffee waste and sewage sludge. Waste Management. 2006;**26**(2):176-181

[63] Juliastuti SR, Widjaja T, Altway A, Iswanto T. Biogas production from pretreated coffee-pulp waste by mixture of cow dung and rumen fluid in co-digestion. AIP Conference Proceedings. 2017;**1840**(May)

[64] Menéndez JA, Domínguez A, Fernández Y, Pis JJ. Evidence of self-gasification during the microwave-induced pyrolysis of coffee hulls. Energy & Fuels. 2007;**21**(1):373-378

[65] Saenger M, Hartge E-U, Werther J, Ogada T, Siagi Z. Combustion of coffee husks. Renewable Energy. 2001;**23**(1):103-121

[66] Sathianarayanan A, Khan AB. An eco-biological approach for resource recycling and pathogen (*Rhizoctonia solani* Kuhn.) suppression. Journal of Environmental Protection Science. 2008;**2**(October):36-39

[67] Kassa H, Workayehu T. Evaluation of some additives on coffee residue (coffee husk and pulp) quality as compost, southern Ethiopia. International Invention Journal of Agricultural and Soil Science. 2014;**2**

[68] Dzung NA, Dzung TT, Khanh VT. Evaluation of coffee husk compost for improving soil fertility and sustainable coffee production in rural Central Highland of Vietnam. Resource Enviroment. 2013;**3**(4):77-82

[69] Brand D, Pandey A, Roussos S, Soccol CR. Biological detoxification of coffee husk by filamentous fungi using a solid state fermentation system. Enzyme and Microbial Technology. 2000;**27**(1-2):127-133

[70] Murthy P, Manonmani M. Bioconversion of coffee industry wastes with white rot fungus *Pleurotus florida*. Research Journal of Environmental Sciences. 2008;**2**(2):145-150

[71] Santos da Silveira J, Durand N, Lacour S, Belleville M-P, Perez A, Loiseau G, et al. Solid-state fermentation as a sustainable method for coffee pulp treatment and production of an extract rich in chlorogenic acids. Food and Bioproducts Processing. 2019;**115**:175-184

[72] Moreira MD, Melo MM, Coimbra JM, dos Reis KC, Schwan RF, Silva CF. Solid coffee waste as alternative to produce carotenoids with antioxidant and antimicrobial activities. Waste Management. 2018;**82**:93-99

[73] Ballesteros-Gómez A, Rubio S. Environment-responsive alkanol-based supramolecular solvents: Characterization and potential as restricted access property and mixed-mode extractants. Analytical Chemistry. 2012;**84**(1):342-349

[74] García-Fonseca S, Ballesteros-Gómez A, Rubio S. Restricted access supramolecular solvents for sample treatment in enzyme-linked immuno-sorbent assay of mycotoxins in food. Analytica Chimica Acta. 2016;**935**:129-135

[75] Accioni F, García-Gómez D, Girela-López E, Rubio S. SUPRAS extraction approach for matrix-independent determination of amphetamine-type stimulants by LC-MS/MS. Talanta. 2018;**182**:574

[76] Ruiz F-J, Rubio S, Pérez-Bendito D. Water-induced coacervation of alkyl carboxylic acid reverse micelles: Phenomenon description and potential for the extraction of organic compounds. Analytical Chemistry. 2007;**79**(19):7473-7484

[77] Mejia EG, Ramirez-Mares MV. Impact of caffeine and coffee on our health. Trends in Endocrinology and Metabolism. 2014;**25**(10):489-492

[78] Golzarand M, Toolabi K, Aghasi M. Effect of green tea, caffeine and capsaicin supplements on the anthropometric indices: A meta-analysis of randomized clinical trials. Journal of Functional Foods. 2018;**46**:320-328

[79] Beyer LA, Hixon ML. Review of animal studies on the cardiovascular effects of caffeine. Food and Chemical Toxicology. 2018;**118**:566-571

[80] Santana-Gálvez J, Cisneros-Zevallos L, Jacobo-Velázquez D, Santana-Gálvez J, Cisneros-Zevallos L, Jacobo-Velázquez DA. Chlorogenic acid: Recent advances on its dual role as a food additive and a nutraceutical against metabolic syndrome. Molecules. 2017;**22**(3):358

[81] Nam S-H, Ko J-A, Jun W, Wee Y-J, Walsh MK, Yang K-Y, et al. Enzymatic synthesis of chlorogenic acid glucoside using dextransucrase and its physical and functional properties. Enzyme and Microbial Technology. 2017;**107**:15-21

[82] Hwang SH, Zuo G, Wang Z, Lim SS. Novel aldose reductase inhibitory and antioxidant chlorogenic acid derivatives obtained by heat treatment of chlorogenic acid and amino acids. Food Chemistry. 2018;**266**:449-457

[83] Muthuswamy S, Rupasinghe HPV. Fruit phenolics as natural antimicrobial agents: Selective antimicrobial activity of catechin, chlorogenic acid and phloridzin. Journal of Food, Agriculture and Environment. 2007:5

[84] Heeger A, Kosińska-Cagnazzo A, Cantergiani E, Andlauer W. Bioactives of coffee cherry pulp and its utilisation for production of cascara beverage. Food Chemistry. 2017;**221**:969-975

[85] Bressani R. Potential uses of coffee-berry by-products. Braham JE, Bressani R, editores. Coffee Pulp: Composition, Technology and Utilization. Ottawa; 1979

[86] de Melo Pereira GV, de Carvalho Neto DP, Magalhães Júnior AI, Vásquez ZS, Medeiros ABP, Vandenberghe LPS, et al. Exploring the impacts of postharvest processing on the aroma formation of coffee beans—A review. Food Chemistry. 2019;**272**:441-452

[87] Komes D, Bušić A. Antioxidants in coffee. In: Processing and Impact on Antioxidants in Beverages. Amsterdam: Elservier; 2014. pp. 25-32

[88] Muñoz W, Chavez W, Pabón LC, Rendón MR, Patricia-Chaparro M, Otálvaro-Álvarez ÁM. Extracción de compuestos fenólicos con actividad antioxidante a partir de Champa (*Campomanesia lineatifolia*). Revista CENIC Ciencias Químicas. 2015;**46**

[89] Echeverria MC, Nuti M. Valorisation of the residues of coffee agro-industry: Perspectives and limitations. The Open Waste Management Journal. 2017;**10**(1):13-22

Chapter 8

Revalorization of Coffee Waste

Felipe J. Cerino-Córdova, Nancy E. Dávila-Guzmán, Azucena M. García León, Jacob J. Salazar-Rabago and Eduardo Soto-Regalado

Abstract

One of the household methods most used to prepare the coffee beverage is the coffee dripping method, which generates millions of tons of coffee waste (CW). Its disposition without control causes environmental matters due to the high consumption of oxygen during its discomposing process. However, the high availability, low cost, and chemical composition of CW (cellulose, hemicelluloses, lignin, ashes, protein, aliphatic acids, fats, and water) make them useful material for obtaining added-value products and bioenergy. In this chapter, the state of the art of different sustainable alternatives to revalorize CW is shown. CW has been successfully applied as an adsorbent for removing pollutants from wastewater and gas, a precursor for obtaining activated carbon, and a feedstock for producing energy and valuable products using mono-process extraction and biorefinery.

Keywords: spent coffee ground, adsorption, biorefinery, bioenergy, activated carbon, experimental design

1. Introduction

Today, coffee is the second worldwide traded commodity after the oil, and it is the second beverage most popular after water. The importance of the global coffee sector can be pointed out due to its presence in 80 countries employing approximately 100 million people [1]. In January 2020, the International Coffee Organization (ICO) estimated that coffee consumption would increase from 1.24 million bags to 169.34 million bags of coffee by the year 2019/2020 [2]. According to these data, there will be a high quantity of spent coffee grounds (SCGs) produced from coffee beverage preparation, which would be released as domestic or industrial trash and cause environmental matters. SCG is considered a toxic residue due to its content of polyphenols, tannins, and caffeine. It has been estimated that 1 ton of green coffee beans can generate 650 kg of SCG, and 1 kg of soluble coffee produced makes 2 kg of SCG wet [3, 4]. The high availability and low cost of SCG allow its revalorization for obtaining valuable products, such as chemical products, activated carbon, biodiesel, and bioenergy.

This chapter will briefly discuss the different ways to revalorize coffee waste. In the first part of this chapter, physicochemical properties are explained since they represent the first stage on SCG revalorization. In the second part, the use of coffee waste as an adsorbent for the removal of pollutants from liquids and gases is shown. The activated carbon produced from coffee waste and its utilization as an adsorbent to remove organic and inorganic pollutants is another topic explored. The recovery of valuable compounds and energy using mono-process extraction and biorefinery from coffee waste will be reviewed. Finally, the experimental design methods to optimize the different processes of coffee waste revalorization are analyzed.

1.1 Physico-chemical properties of coffee waste

The biomass revalorization, such as coffee waste, depends primarily on their physicochemical properties, such as chemical composition, presence of extractable compounds, and diversity of functional groups. These properties are altered according to the type and plant variety; in the case of coffee, the most commonly used is the so-called Arabica coffee, so their main physicochemical characteristics were briefly analyzed.

1.1.1 Chemical composition

Coffee waste, being lignocellulosic biomass, which is mainly composed of the essential life elements (C, H, O, and N), which are primarily forming cellulose (59.2–62.94 wt%), hemicellulose (5–10 wt%), and lignin (19.8–26.5 wt%) [5, 6]. Besides, these elements are present in the form of recoverable compounds, such as essential oils and flavonoids, among others. However, since this material has already been subjected to a hydrothermal extraction process, the presence of these compounds is usually low compared with lignocellulosic constituents (10 wt%) [6]. Moreover, this type of waste usually has some elements considered inorganic micronutrients such as calcium, magnesium, or sodium, but their concentrations are generally less than 5.0% dry weight [5–7].

The main component of plant biomass is cellulose, which is made up of linear chains of D-glucose linked by β-1,4 bonds, and it has a form of crystalline fibrillar aggregates, which are formed due to the hydrogen bonds among the HOS present in the D-glucose, as can be seen in **Figure 1**. On the other hand, hemicellulose forms an aggregate of simple sugars of different structures that are attached to cellulose microfibers. Several authors had reported the presence of xylose, arabinose, galactose, and mannose in coffee residues. These types of molecules usually present cyclic structures of 5 or 6 constituents, being abundant in alcohol groups. However, their heterogeneity makes impossible the formation of crystalline arrangements [7, 8]. On the other hand, lignin, whose molecular representation is illustrated in **Figure 2**, is a biopolymer, not a polysaccharide, which is considered the most abundant in plant biomass. This biopolymer has a high structural diversity originated from the enzymatic dehydrogenation of coumaryl, coniferyl, and sinapyl alcohols and subsequent radical polymerization. This heterostructure provides properties such as hardness, resistance to microbial attacks, and oxidative stress, complicating its biodegradation [9].

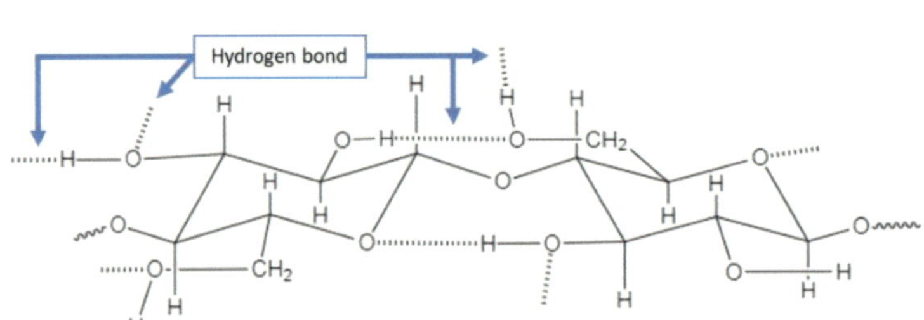

Figure 1.
Cellulose structure showing the hydrogen bonds.

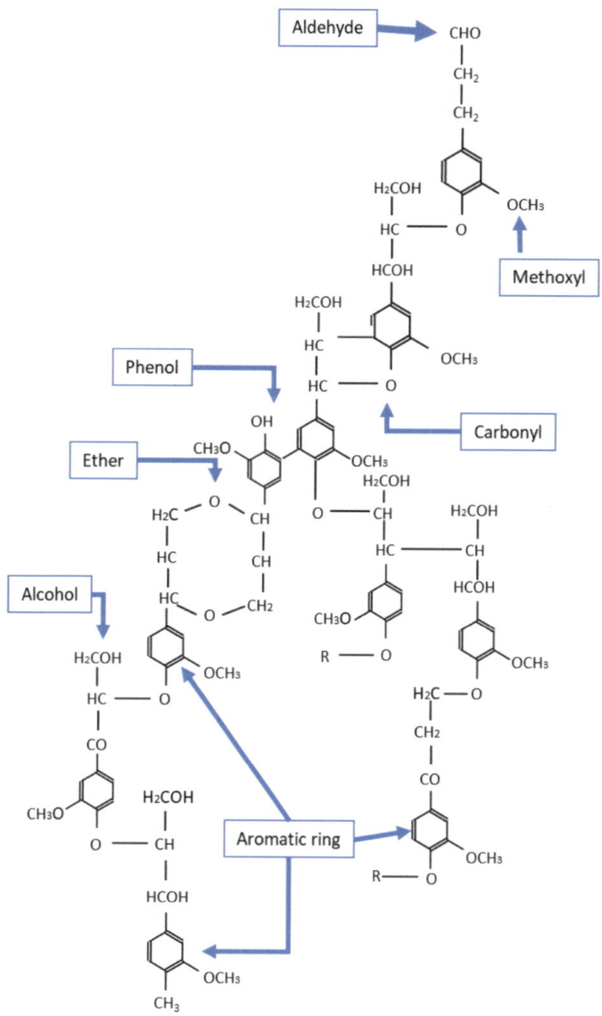

Figure 2.
Lignin chemical structure.

1.1.2 Surface physicochemical properties

Given the structural diversity of the constituents of the coffee residue, a heterogeneous presence of functional groups on the surface of the material is expected, which will provide this biomass with unique characteristics. Cellulose and hemicellulose have functional groups of the alcohol type (R—OH), which can favor the functionalization of these materials, for example, through esterification processes [9]. On the other hand, given its formation process, lignin as a macromolecule has phenolic and aliphatic hydroxyl groups, in addition to methoxyl, carbonyl, and aldehyde groups, among others [8]. The concentration of these groups will depend on the variety and class of the starting material. The structure of lignin is shown in **Figure 2**, and the functional groups mentioned above are indicated; it is important to highlight that lignin has aromatic rings capable of promoting interactions π-π^* with other compounds, which could allow the use of coffee waste as an adsorbent for organic compounds [10].

Among the various analytical techniques used to characterize solid materials is infrared spectroscopy with Fourier transform, which allows identifying surface functional groups simply and effectively. The infrared spectrum of coffee waste is presented in **Figure 3**. In it, the wavelengths at which the various vibrational modes

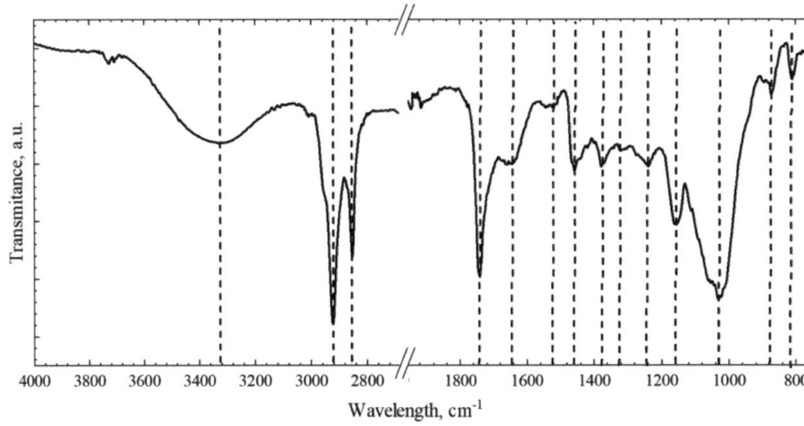

Figure 3.
FTIR spectrum of a sample of coffee waste.

of the surface groups can be detected are indicated. The absorption bands found are similar to those reported by multiple authors for coffee residues of the Arabica variety [5, 6, 8]. In the spectrum, two absorption regions can be evidenced, the first one from 3800 to 2700 cm^{-1}, finding signals around 3340 cm^{-1} corresponding to the vibrations of the OH bonds present in the alcohol groups, followed by a doublet of bands at 2920 and 2860 cm^{-1} of the CH interactions, present in all lignocellulosic structures. The second region, between 1900 and 750 cm^{-1}, has a higher number of corresponding bands with links C=O of the carbonyl groups present in the aldehydes (1740 cm^{-1}); C=C of the double bonds of the aromatic structures of lignin (1640, 1525, and 1475 cm^{-1}); CH of the methyl and methylene groups of the polymer chains of the constituents (1440 and 1380 cm^{-1}); CO of the groups of the ester type (1320, 1240, and 1160 cm^{-1}) and alcohol (1030 cm^{-1}); and finally, the bands located at 870 and 810 cm^{-1} are characteristic signs of substitutions in aromatic structures. Together these bands corroborate the polymeric nature of the coffee residue and make it possible to elucidate, at least qualitatively, the type of surface structures it possesses. The functional groups detected on the surface of the material are primarily acidic, which means that they are capable of yielding the proton and therefore can grant a negative charge density to the biomass surface depending on the pH of the medium. Volesky [11] and Ahsan et al. [12] reported that this type of functional group acts as active sites in the processes of pollutant removal. Several studies have quantified the presence of this type of active sites, indicating in a general way the predominance of phenolic, carbonyl, and carboxylic sites [5, 9, 13].

2. Pollutants removal by coffee waste

Coffee wastes in their several forms (e.g., coffee husks, coffee silverskin, coffee bean skins, and spent coffee grounds) have been used in the removal of inorganic and organic compounds from aqueous solutions at least for the last two decades. The first report about the use of coffee wastes for the removal of pollutants from wastewater was published in 2002 [14]. In this study, the authors evaluated several adsorbents, including coffee bean skins (CBSs), for the removal of copper and zinc ions from swine breeding wastewater. The copper removal efficiency of CBS was about 50%, whereas no zinc adsorption was obtained. However, no insight regarding the adsorption mechanism was provided. An attempt to elucidate the adsorption mechanism of metal ions was made by measuring the isotherms of lead adsorption onto degreased and protein-denatured coffee grounds [15]. The amount of lead adsorption onto degreased coffee grounds did not exhibit significant change compared to that on coffee grounds. On the contrary, protein-denatured coffee grounds had an adverse effect on the amount of lead adsorbed. These results indicate that fat cannot adsorb lead ions, but proteins contained in coffee grounds are responsible for the removal of lead ions. Also, it was demonstrated that there is no dependence on the type of coffee beans (e.g., *C. robusta*, *C. arabica* from four different regions) in the adsorption of lead ions due to their similar protein content.

Untreated coffee husks (UCH) have been successfully used for the removal of several heavy metal ions such as chromium (Cr), copper (Cu), cadmium (Cd), and zinc (Zn). Oliveira et al. [16] reported maximum adsorption capacities of 7.5, 6.96, 6.85, and 5.56 mg/g for the adsorption of Cu, Cr, Cd, and Zn onto UCH, respectively. In this study, Boehm titration was used to determine the functional groups before and after the adsorption experiments. The authors observed a decrease in the quantity of functional groups due to heavy metals' adsorption. The results showed that all functional groups (carboxylic, lactonic, phenolic, and basic groups) were involved in the adsorption of heavy metal ions, with relative affinities as follows: Cu > Cr > Cd > Zn for basic groups; Zn > Cu > Cr > Cd for carboxylic groups; Cr > Zn > Cd > Cu for lactonic groups; and Cr > Zn > Cu > Cd for phenolic groups.

Coffee silverskin (CS) is another relevant coffee waste evaluated for the removal of metal ions. CS is part of the outer layer of green coffee beans, which is generated during the roasting process, and it has no commercial value [17]. CS demonstrated similar adsorption efficiency of Ni and Zn when it was compared to SCG, while Cu ions were removed to a lesser extent by using CS. The authors attributed the higher performance of SCG to the higher content of lignocellulosic components. The maximum adsorption capacities on CS were 15.17, 9.58, and 1.43 mg/g, respectively, for Cu, Zn, and Ni ions.

Among the different forms of coffee wastes, spent coffee grounds (SCGs) collected from coffee shops or cafeterias have become one of the most popular coffee wastes studied for the removal of pollutants. Azouaou et al. [18] used them without treatment for the removal of Cd ions from aqueous solution. The authors reported an adsorption capacity of 15.65 mg/g and 120 min to achieve the adsorption equilibrium. Also, it was demonstrated that the particle size does not affect the removal of Cd, suggesting that intraparticle diffusion is not the rate-limiting step. Davila et al. investigated the adsorption mechanism of copper ions onto SCG [6]. They found that the amount of calcium ions and hydrogen ions, released from SCG carboxyl and hydroxyl groups during Cu adsorption, were similar to the amount of Cu ions adsorbed. Thus, the adsorption of Cu ions onto SCG was mainly due to ion exchange. The maximum adsorption capacity obtained was 14 mg/g. Similarly, Gomez-Gonzalez et al. [13] conducted the adsorption of Pb ions by SCG and examined the pH effect on the adsorption capacity. An increase in pH caused an increment of the adsorption capacity of Pb, and the maximum adsorption capacity reported was 22. 9 mg/g at pH 5. On the other hand, Elsherif et al. [19] evaluated the removal of cobalt by SCG. The authors reported a maximum adsorption capacity of 243.9 mg/g.

Additionally, SCG has been used for the simultaneous removal of metal ions from aqueous solutions. In this regard, Futalan et al. [20] evaluated the performance of SCG for the simultaneous removal of Cu, Pb, and Zn from soil washing wastewater. The maximum removal efficiency obtained was 57.23, 68.73, and 84.55% for Pb, Cu, and Zn ions, respectively. The removal of mercury ions by SCG was reported by Mora Alvarez et al., and the maximum adsorption capacity was found to be 31.75 mg/g [21]. Two desorption agents were evaluated, nitric acid and chloride acid, where the latter presented better desorption of Hg ions. However, when SCG was subjected to one adsorption-desorption cycle, a loss of removal efficiency was observed, decreasing from 97 to 28% Hg removal. On the contrary, Kyzas [22] demonstrated the strong reuse potential of SCG in the adsorption of Cu and Cr ions since only 10% of metal ion uptake was loss after 10 cycles of adsorption-desorption. Similarly, the adsorption capacity of Cu, Cd, and Pb ions by SCG remained the same during four adsorption-desorption cycles according to the report by Davila et al. [23]. In this study, SCG regeneration was carried out using citric acid, calcium chloride, and nitric acid as eluent agents. The trend of the desorption efficiency through four adsorption-desorption cycles was $HNO_3 > CaCl_2 > C_6H_8O_7$.

Although most applications of coffee waste have been made for the removal of inorganic pollutants from water, coffee waste also has demonstrated the potential for the removal of organic pollutants. For example, methylene blue (MB) was removed from aqueous solutions by UCH [24]. The results showed that above the point of zero charge of the UCH (approx. pH 4.5), there was no pH effect on the removal of MB. The maximum adsorption capacity of MB onto UCH was 55.3 mg/g. MB has been used as a model dye molecule to demonstrate the potential of an adsorbent for the removal of dyes from wastewater. The capability of coffee waste for the removal of organic pollutants is associated with the density of the oxygen-containing functional groups that increase the p-p interaction force between the coffee wastes and the organic molecules. Accordingly, Dai et al. [25] proposed an adsorption mechanism for tetracycline (typical bactericidal drug) onto SCG by pi-pi interaction between the aromatic ring of the tetracycline molecule (TC) and the aromatic functional groups of the SCG. The maximum adsorption capacity of TC onto SCG was found to be 64.89 mg/g. Also, the effect of ionic strength on TC adsorption was evaluated, where there was a competition for the adsorption sites, diminishing the adsorption capacity as the ionic strength was augmented.

It is noteworthy mentioning that most of the studies on pollutant removal by coffee wastes have been carried out in batch configuration. However, adsorption by continuous fixed bed systems are the common configuration used in industrial applications due to the high volume of pollutant-solution processed, operation simplicity, and higher mass transfer characteristics than batch systems. Despite that, only a few reports have been made on the use of coffee waste in fixed-bed columns. Utomo et al. [26] conducted column adsorption experiments for the removal of Cu, Zn, Cd, and Pb ions by SCG. The adsorption efficiencies were higher than 91% for all metal ions. Besides, the percentage of Cu ions adsorbed by a column packed with SCG was shown, where it can observe the breakthrough at 100 mL (30 min). A thorough study of the performance of Cd, Cu, and Pb ion removal in a fixed-bed column packed with SCG was presented by Davila et al. [23]. The effect of the process variables (e.g., flow rate, bed height) was evaluated, and the maximum breakthrough times of Cd, Cu, and Pb ions were 50, 160, and 220 min, respectively. Furthermore, the breakthrough curves were predicted well by using a mass transfer model that includes axial dispersion, external mass transfer resistance, and ion-exchange model to describe the equilibrium adsorption.

All the applications mentioned above of coffee wastes were about the removal of inorganic or organic pollutants from wastewater. Only one study has reported the use of coffee wastes for the removal of a gaseous pollutant [27]. In this study, a decrease in 43% on the ozone concentration in an ozone-filled chamber was achieved by using SCG, which was competitive to the performance of commercial activated carbon (about 56%).

3. Pollutants removal by modified coffee waste

High volumes of coffee waste with no commercial value are generated worldwide daily, causing an environmental burden. For this reason, several studies have been conducted to reuse coffee wastes as adsorbents for the removal of several pollutants. Although untreated coffee wastes have demonstrated adsorption capacities similar or even higher than those obtained by commercial materials (e.g., activated carbon), recent studies have focused on the modification of coffee wastes to increase even further the removal efficiency. In this sense, Lafi et al. [28] modified the surface of commercial coffee waste with cationic surfactants, cetyltrimethylammonium bromide (CTAB) or cetylpyridinium chloride (CPC), to increase the affinity for the anionic dyes such as methyl orange (MO). The maximum adsorption capacity obtained for MO was 58.82 and 62.5 mg/g, onto CTAB-coffee waste and CPC-coffee waste, respectively. On the other hand, Cerino-Córdova et al. [29] modified the surface of SCG with citric acid to increase the amount of carboxylic groups. By doing that, the adsorption capacity of Pb and Cu ions was 3.2 and 8.1 times higher than those obtained by the unmodified SCG. Similarly, Botello-Gonzalez et al. [9] investigated the adsorption capacity of SCG modified with citric acid in the competitive adsorption of Pb and Cu ions. The maximum adsorption capacities of Pb and Cu ions were 130 and 45.4 mg/g, respectively. Additionally, the authors proposed a model based on ion exchange that takes into account the surface chemistry of the modified SCG interaction with the heavy metal ions in the liquid phase. In another study, SCG was acid activated with hydrochloric acid and examined for the removal of lead and fluoride ions [30]. The maximum adsorption capacities were 65.4 and 9.75 mg/g of Pb and F ions, respectively. Another acid activation of the surface groups of coffee waste was carried out with sulfuric acid [31]. The sulfonate coffee waste (CW-SO_3H) was successfully used for the removal of bisphenol A (BPA) and sulfamethoxazole (SMX) from water. Highly negative surface charge was obtained after the incorporation of the sulfonic acid groups, increasing the interaction with the cationic pollutants. The maximum adsorption capacities were found to be 271 and 256 mg/g for BPA and SMX, respectively. Besides chemical modification of coffee waste, physical activation has been employed successfully for the removal of metal ions. For example, Delil et al. [32] conducted the reduction of the grain size of SCG by an ultrasonic process to increase the specific surface area. Also, the zeta potential of the activated SCG was more negative after the ultrasonic method, enhancing the adsorption of cadmium ions.

4. Pollutant removal by coffee waste composites

Composite adsorbents with coffee waste (CWC) have been synthesized and examined for the removal of pollutants from aqueous solutions. In this regard, some studies have evaluated the encapsulation of coffee wastes in polysaccharides such as calcium alginate (CA) and chitosan (Cs). For instance, spent coffee grounds were encapsulated by using CA to increase the adsorption capacity of Ni, Cd, and Cu [33, 34]. The results showed high adsorption capacities and faster adsorption rates than CA beads alone. In another study, coffee wastes were mixed with Cs and poly(vinyl alcohol) (PVA) to enhance the adsorption capacity of pharmaceuticals [35]. The addition of coffee wastes to the matrix of Cs-PVA allowed an increase in the adsorption of metamizole (MET), acetylsalicylic acid (ASA), acetaminophen (ACE), and caffeine (CAF) as compared to the virgin material.

On the other hand, coffee waste composites with magnetic properties have been synthesized to facilitate the removal of the adsorbent from the liquid media. In this sense, magnetic coffee waste composite prepared from Fe_3O_4, PVA, and alkaline pretreated SCG was evaluated for the removal of Pb ions from aqueous solutions [36]. The maximum adsorption capacity of Fe_3O_4/PVA/APSCG of Pb ions was reported as 57 mg/g. Similarly, a magnetic coffee waste composite was prepared by using SCG and Fe_3O_4, without PVA as a cross-linking agent [37]. The maximum adsorption capacity of Pb ions was found to be 41.15 mg/g when a 2% loading of Fe_3O_4 nanoparticles was used. A further increase in the Fe loading decreased the removal of Pb ions due to the agglomeration of Fe_3O_4 on SCG.

Other types of coffee waste composites studied for the removal of pollutants from aqueous solutions are those obtained from the combination of clay or siliceous materials with coffee waste. In this regard, limestone combined with SCG was synthesized for the removal of both anionic and cationic dyes (methylene blue and orange II, respectively) [38]. The maximum removal percentage for methylene blue (MB) and orange II (OR II) was 100 and 85% at pH 8 and 2, respectively. However, in competitive adsorption experiments, the presence of MB causes a reduction in the removal of OR II from 85 to 60%. Another coffee waste composite reported as a heavy metal scavenger is composed of coffee wastes and attapulgite clay (SCG-AC) [39]. The maximum adsorption capacity of Pb ions was reported to be 4.45 mg/g.

5. Solid coffee waste as a precursor to activated carbon

The use of lignocellulosic waste to obtain valuable products has been proved to be an critical ecological strategy because these wastes are widely available. Therefore, these wastes represent a pollution problem in the water, soil, and air. A pyrolysis process can be used to obtain some valuable products, such as biofuels and activated carbon, among other useful products. Activated carbon is widely used as an adsorbent material to remove pollutants from aqueous solutions and to capture CO_2 or H_2S in the gas phase. Thus, by using lignocellulosic waste, it is possible to prevent soil, water, and air pollution and to apply activated carbons in tertiary treatment of wastewater.

5.1 Pyrolysis process

There are many sources to obtain agricultural waste, for instance: barley husks, coconut shells, sawdust, and spent coffee grounds, among others. These wastes have different percentages of cellulose, hemicellulose, and lignin. Today, agricultural wastes are readily available and are released to the environment or used for other proposes, such as livestock feed. The content of fixed carbon in these wastes and their abundance has led several researchers to investigate the use of these wastes as precursors to produce activated carbon, which can be used as adsorbent material to remove pollutants from aqueous solutions.

The pyrolysis process is useful to obtain some valuable products from lignocellulosic biomass. Pyrolysis means the thermal decomposition of lignocellulosic biomass under an inert atmosphere, for instance: nitrogen, argon, steam, and carbon dioxide, among others. The products of the pyrolysis process include biochar, biofuel, and volatile compounds. To determine the appropriate temperature range to carry out this process, a thermogravimetric analysis is required. Thus, the process is usually performed within temperature ranges from 400 to 600°C and from 700 to 1200°C for chemical and physical activations, respectively. During the pyrolysis process, the biomass loses humidity between 100 and 200°C. At temperatures higher than 200°C, cellulose, hemicellulose, and lignin contents are decomposed at different temperature ranges, besides volatile compounds are released, which content condensable vapors (phenol and aromatics, among others), and light hydrocarbon compounds.

The pyrolysis mechanism of lignin is more complex than that of cellulose and hemicellulose. During the lignin decomposition, there are primary reactions in the range of 200–400°C and secondary reactions at temperatures higher than 400°C [40]. On the other hand, at temperatures of 200–400°C, hemicellulose is broken down [41], and cellulose can be decomposed in a temperature range of

315–400°C [42]. All these decomposition processes lead to polymerize pyrolytic products to develop activated carbons.

5.2 Biochar activation

The biochar obtained in the pyrolysis process can be subjected to an activation process, which is a method useful to develop the physical and textural properties of the adsorbent material, such as total pore volume, surface area, and porosity. Besides, the activation process widens the pore diameter from nanopores to mesopores and macropores. This improves the internal diffusion of the pollutants inside the adsorbent particle. The activation of carbon can be carried out by physical or chemical activation. Chemical activation can be performed at a temperature range of 400–700°C by using inorganic compounds. On the other hand, the temperature range for physical activation with steam or CO_2 is from 700 to 1200°C, which means more power consumption.

5.3 SCG activated carbon

Commercial activated carbon (CAC) is a useful material to remove pollutants from aqueous solutions. However, CAC can be expensive; for this reason, some researchers have studied several materials to produce activated carbon from lignocellulosic wastes by pyrolysis such as coconut shell, corncob, carnauba pall and fine nut, sawdust, and candied chestnut [43–47]. Given the lignocellulosic structural nature of solid coffee residues, carbon content is predominant compared to other constituent elements. This, along with the abundance of the residue, makes it an optimal material as a precursor in the synthesis of activated carbon [48, 49].

Several researchers have reported the use of coffee waste to produce activated carbon. **Table 1** shows the activation conditions to produce SCG activated carbon by chemical or physical activation. SCGs were chemically activated by KHO, $ZnCl_2$, H_3PO_4, or H_2SO_4 or physically activated. The activation temperatures were between 400 and 800°C, and it is important to mention that chemical activation allows low temperatures for the pyrolysis process, instead of physical activation. In most cases, the yield and pore size were reported. A high pore diameter is an important parameter because it allows for the internal diffusion of pollutants inside the adsorbent particle, enhancing the adsorption capacities. According to the data shown in **Table 1**, SCG is a viable option to produce activated carbon because the obtained SCG carbon has a high surface area and a reasonable pore width, which are relevant parameters to carry out an adsorption process to remove pollutants from aqueous solutions.

The use of coffee extract residue to produce ethanol and activated carbon was conducted and studied by Fotouhi et al. [46]. The coffee solid residue was chemically activated by using H_3PO_4 at 600°C and physically activated with steam at 700°C. The produced adsorbent showed a pore volume range from 0.22 to 0.59 cm^3/g and a surface area from 233 to 696 m^2/g. Diaz de León et al. [50] reported the use of SCG to produce activated carbon by chemical activation with $ZnCl_2$. An experimental design was carried out varying three factors: temperature (450–600°C), activation time (40–120 min), and impregnation ratio mass of $ZnCl_2$: the mass of spent coffee ground (0.5:1.5). The optimal conditions reported were 600°C, 40 min of activation time, and 1.5 g $ZnCl_2$/g SCG. At these conditions, a surface area of 1280 m^2/g, a yield of 26%, and a total pore volume of 0.77 cm^3/g were reported. The adsorbent obtained was used to remove phenol from aqueous solutions at pH 7, and maximal adsorption capacity of 160.52 mg of phenol/g was reached.

Surface area (m^2/g)	Total pore volume (cm^3/g)	Pore width (nm)	Pyrolysis temperature (°C)	Activation agent	Flow	Yield (%)	Ref.
1040.3	0.635	—	700–900	KOH	Ar	—	[48]
1039	0.481	4.7	500	$ZnCl_2$ Steam	N_2	40% 20%	[49]
1280	0.77	3	600	$ZnCl_2$	N_2	26	[50]
831	0.44		400, 450, and 500	$ZnCl_2$	Air	15.99 22.95%	[51]
1121	0.954	1–3	800	$ZnCl_2$	N_2	—	[52]
2785	1.36	1.051	400 and 700	KOH	N_2	11–16%	[53]
1778	0.657	—	800	KOH	N_2	—	[54]
146.1	0.0705	1.6	600	H_2SO_4	N_2	42.77–51.85%	[55]
1082	0.51	3.0	600 and 700	KOH or CO_2	N_2	23–29%	[56]
889 ($ZnCl_2$) 1003 (HPO_3)	0.765 ($ZnCl_2$) 0.618 (HPO_3)	3.44 ($ZnCl_2$) 2.44 (HPO_3)	600	$ZnCl_2$ or H_3PO_4	N_2	—	[57]

Table 1.
Reports of chemical activation conditions to produce activated carbon from coffee wastes.

According to the data shown in **Table 1**, SCG can be considered an excellent precursor to produce activated carbon. The large surface area achieved in SCG activated carbon could be used to remove inorganic and organic compounds from aqueous solutions.

5.3.1 Removal organic and inorganic pollutants by activated carbon-spent coffee grounds

Activated carbon is a material in which carbon is forming disordered graphite plates, in whose peripheries there is a wide diversity of functional groups, which gives it unique physico-chemical properties. Additionally, this material usually has raised surface areas, generally greater than 1000 m^2/g, which develops through various oxidation reactions [58]. Given these characteristics, this material is typically used in numerous applications, excelling in the removal of organic and inorganic compounds present in the gas and liquid phase.

Waste coffee grounds were used to produce activated carbon by KOH under Ar atmosphere at three temperatures (700, 800, and 900°C), the adsorbent material was tested to adsorb CH_4 and H_2. The activated carbon at 900°C showed a CH_4 adsorption capacity of 1.96 mmol/g at 273 K and 100 kPa. However, at 3000 kPa, the highest adsorption capacity was reported to be 4.2 mmol/g. The three adsorbents materials (700, 800, and 900°C) were also tried to adsorb H_2 at 77 K and 100 kPa, achieving the highest adsorption capacity of 1.75 wt% [48]. Activated carbon from waste SCG as the precursor was physically activated by CO_2 or steam at high temperatures (700–900°C) and chemically activated by $ZnCl_2$, KOH, and H_3PO_4 at 450 and 600°C. Nevertheless, in this work, only the raw material, the activated carbons by $ZnCl_2$ or steam, was tested to adsorb Bisphenol-A. The removal of Bisphenol-A was found to be 98, 12, and 0% for carbon activated by $ZnCl_2$, raw material, and carbon activated by steam, respectively. These results were compared with a commercial activated carbon, which showed a Bisphenol-A removal of 93%. The poor adsorption performance of the SCG carbon activated by steam is due to the low surface area reported for this material (4 m^2/g) [46].

SCG was used as a precursor to prepare activated carbon by chemical activation with $ZnCl_2$ at three impregnation ratios, at room temperature, and during 8, 12, and 24 h. The adsorption of Cu(II) was conducted using this activated carbon. The experimental data were fitted by using the Langmuir, Freundlich, and Elovich isotherms; the maximum adsorption capacity (Langmuir) was 285.71 mg/g, and the maximum Cu(II) removal reported was 18% at 100 rpm, and the pH solution value was not reported [51].

The production of SCG-based activated carbon was carried out at three impregnation ratios, g $ZnCl_2$/g precursor (1:0.5, 1:1, and 1:2); the impregnated precursor was carbonized under N_2 atmosphere at 800°C during 60 min, and this material was tested for H_2S separation. The SCG activated carbon was used to study H_2S dynamic breakthrough capacity passing a dilute flow of H_2S-Air (1000 ppm, 80% humidity) through a fixed bed. The adsorbents activated at impregnation ratios of 1:2 and 1:1 showed the lowest (18.2 mg/g) and the highest breakthrough capacity (127 mg/g), respectively [51].

Chemical activation of SCG was performed with KOH using 2:1 and 4:1 impregnation ratio (KOH: precursor) to produce activated carbons. The carbonization process was carried out at 400 or 700°C under N_2 flow for 2 h. These activated carbons were tested to adsorb CO_2 at 0, 25, and 50°C and 0–10 bars. The highest adsorption capacity obtained was 6.8 mmol/g at 1 bar and 0°C, when the activated carbon was produced at 700°C and an impregnation ratio of 4:1. However, when the carbonization process was performed at 10 bars, the highest uptake achieved was 23.26 mmol/g [53].

SCG microporous activated carbon (using potassium hydroxide as activation agent) was synthesized and characterized. The precursor was pyrolyzed using 1:9, 1:18, and 1:36 mmol KOH:g SCG of impregnation ratios and under N_2 flow at 800°C during 1 h. SCG activated carbons were used to adsorb phenol and methylene blue. The equilibrium was attained within 100 and 360 min for phenol and methylene blue, respectively. The maximal adsorption capacity, based on Langmuir isotherm, for phenol and methylene blue was 3008 and 1058 mmol/g, respectively [54].

The influence of the impregnation ratio of H_2SO_4 over SCG granular activated carbon to treat leachate was studied. Six samples of leachate with the following chemical and biological parameters were treated: COD (1010–1815 mg/L), BOD_5 (184–338 mg/L), NH_4-N (2208–2780 mg/L), iron (4.25–4.73 mg/L), and PO_4-P (220–284 mg/L). However, in this research, only the removal percentage of iron and PO_4-P was found to be 77 and 84%, respectively, when impregnation ratios of 2.5 and 0.5 were used [55].

SCG obtained from a trademark coffee (Nespresso®) was used as a precursor to produce activated carbon. KOH was used as an activating agent at four impregnation ratios. The chemical activation process was carried out at 873 K, and physical activation with CO_2 was carried out at 973 K. These materials were used to capture CO_2, which is a byproduct of the combustion process. However, in this study, a pure CO_2 flow was tested at 298 K and 101 kPa. The CO_2 adsorption capacities of the adsorbents activated by using chemical activation and physical activation were 3 and 2.3 mmol/g, respectively [56].

SCG carbon activated with phosphoric acid and zinc chloride was used to adsorb Pb(II) and Cd(II). The precursor was mixed with $ZnCl_2$ or H_3PO_4 at chemical agent/coffee residue mass ratios of 25, 50, 75, and 100% at 85°C for 7 h. When the precursor was activated with H_3PO_4 (50% impregnation ratio), the maximal adsorption capacities, based on Langmuir isotherm, were as follows: 89.28 mg Pb(II)/g and 46.95 mg Cd(II)/g. With $ZnCl_2$ (75% impregnation ratio) as the activation agent, the maximal adsorption capacities were 63.29 mg Pb(II)/g and 37.04 mg Cd(II)/g [57].

The use of activated carbon derived from SCG was recently reported, and the activation procedure was carried out with $ZnCl_2$. To optimize the activated carbon production, an experimental design was performed; the independent factors were temperature (450 and 600°C), activation time (40 and 120 min), and impregnation ratio (0.5 and 1.5 g $ZnCl_2$/g SCG), and the experimental responses were surface area, yield, and hardness. The optimal conditions were impregnation ratios of 1.5, 600°C, and 40 min. At these conditions, the experimental responses were surface area 1279.96 m^2/g, yield 26%, and hardness 76.77%. The activated carbon produced at these conditions was used to adsorb phenol from aqueous solutions, based on Langmuir isotherm, the maximum adsorption capacity was 160.52 mg/g, and the equilibrium was attained less than 150 min [50].

6. Mono-process extraction, bioenergy, and biorefinery

The circular economy demands the efficient utilization of resources in the production systems and the long-term material use by recycling or remanufacturing [59]. This concept can be correctly applied to the product obtained from biomass processing, such as coffee waste. This material could be a feedstock for a mono-process extraction, bioenergy production, and biorefining. The first stage of the process design is determining the composition of SCG, which has been shown to be highly dependent on coffee varieties [60–62]. The range of the biochemical composition values obtained is shown in **Table 2**. It is important to consider that SCG has a high quantity of organic compounds such as polyphenols, polysaccharides, amino acids, fatty acids, and minerals.

6.1 Valuable chemical compound recovery by mono-process extraction

These techniques use chemicals to extract valuable organic compounds (lipids, polysaccharides, phenolics, tannins, and caffeine), and it could be assisted by ultrasound, enzyme, or microwave. These chemical compounds can be useful to obtain high added value products: biodiesel, cosmetics, food additives, pharmaceuticals, packing materials, and adhesives. These techniques are divided into conventional (Soxhlet extraction, maceration, and hydrodistillation) and nonconventional techniques (supercritical fluid extraction, enzyme-assisted extraction, ultrasound-assisted extraction, microwave-assisted extraction, pulsed electric-assisted extraction, and pressurized liquid extraction). The factors studied in the SCG mono-process extraction have been the coffee varieties, solvent, time, pressure, and temperature. The Soxhlet extraction process has several disadvantages, such as low productivity, high solvent consumption, and high extraction time [63]. Ultimately, the main goal of nonconventional methods is to decrease the utilization of synthetic and organic chemicals and operational condition and to improve the yield and quality of extract, which makes them environmentally friendly compared to conventional techniques [64]. Mono-process extraction has been used in SCG for oil, caffeine, phenolic compounds, polysaccharides, and tannin recovering.

6.1.1 Oil recovery

The oil content in SCG is highly dependent on the coffee variety (**Table 2**). It has been demonstrated that oil extracted from SCG could be used in biodiesel production. Solvent extraction and supercritical fluid extraction with CO_2 and ethanol as a solvent

Biochemical composition	
Compound	**Concentration (wt%db*)**
Lipids	6.7–19
Carbohydrates	14.1–72.4
Proteins	4.3–17
Mannose	21.2–47
Galactose	25–30
Glucose	19–24
Arabinose	3.8–6
Caffeine	0.96–7.9
Oil	10–20 wt%
Ultimate analysis	
Element	ww%db*
C	52.1–53
H	6.8–7.03
N	1.71–3.47
S	0.1
O	34.7–38.1
Proximate analysis	
	ww%db*
Moisture	11.5–61
Volatile	79.5
Ash	0.68–2.2
Fixed carbon	8.2

**db: Dry base.*

Table 2.
Composition, ultimate, and proximate analysis of SCG.

have been the methods used for oil recovery. The solvent type (ethanol anhydrous, ethanol hydrous, hexane, and methanol), g SCG/g solvent ratio (20.3–23.8 g/g), extraction time (19.5–30.4 min), and temperature (30°C) were studied. The oil yield obtained varied from 7.5 to 14.7 w/w% d.b. The best oil yield obtained (14.7 w/w% d.b.) was using 22.5 g SCG/g hexane, 30°C, and 30.4 min of extraction time [60, 63, 65].

6.1.2 Caffeine recovery

Caffeine is an alkaloid, which is the coffee chemical compound most recognized in the world. The content of caffeine in coffee beans is higher than SCG; however, a high quantity remains in SCG. The Soxhlet extraction, ultrasound-assisted extraction, membrane technology, and pressurized liquid membrane with ethanol and water have been the methods used for caffeine recovery. The range of caffeine yield was similar for the different methods, falling in the range from 0.734 to 43 mg/g db [60, 61]. However, the pressurized liquid extraction (PLE) has the advantage of decreasing solvent use and operating time, being an oxygen and light-free environment process.

6.1.3 Phenolic compound recovery

SCG has a high content of phenolic compounds (caffeoylquinic, feruloylquinic, p-coumaroylquinic, ferulic, and quinic acids). These have anticancer, antidiabetic, antioxidant, antiviral, antiallergen, antimicrobial, and antifatigue activities. Additionally, these chemical compounds could be incorporated into skincare products. Different methods, such as subcritical water, ultrasound-assisted, pressurized liquid extractions, and supercritical fluid extraction with CO_2, have been used for phenolic compound recovery [61, 62]. The experimental results showed a range of phenolic compound recovery from 19 to 273.4 mg GAE/g. The results demonstrated that the ethanol extraction method with oil extraction by hexane pretreatment was the best process, followed by the autohydrolysis process (273.4 mg GAE/g). The optimal experimental conditions were 5 ml ethanol/g SCG and ambient temperature [60, 61].

6.1.4 Polysaccharides recovery

The polysaccharides in SCG present different structures, such as galactomannans, arabinogalactans, and cellulose, which are used as dietary fiber ingredient in functional food. These compounds have immunostimulatory, antimicrobial, and antioxidant activities. Furthermore, they have good thermal stability properties. Various methods for polysaccharide purification from SCG have been utilized successfully, such as extraction with chemical agents (potassium hydroxide and sulfuric acid), subcritical water hydrolysis, autohydrolysis, and microwave superheated water extraction methods. The polysaccharides extracted from SCG varied from 22 to 61.9 w/w% d.b., and several studies have demonstrated that the yield increases when the coffee is roasted [5, 62, 65–68]. The best method of polysaccharide extraction (61.9 w/w% d.b.) was the microwave superheated water extraction, with the following experimental conditions: 1 g SCG/10 ml of water, 2 min of extraction time, and 200°C.

6.1.5 Tannin recovery

Tannins are low-cost natural biopolymers that could serve as biosorbents and prepare as adhesives. The extraction of tannins from SCG has been carried out by Soxhlet extraction with 5% of sodium hydroxide. The best tannin extraction yield was 21.02 mg tannins/g d.b. at 8.2 g SCG/g NaOH, 30 min of extraction time, and 100°C [69].

6.2 Energy recovery from SCG

The chemical composition of SCG makes them a viable material to use them as feedstock to produce biodiesel, bio-oil, syngas, and energy via a combustion process.

6.2.1 SCG pellets for energy production

The combustion is the process used for obtained energy from SCG due to its calorific value. The SCG can be used after oil and lipid extraction processes. Some studies have been carried out to increase the calorific value of SCG. These wastes have been blended with other materials such as sawdust, beechwood, and glycerol. The solid fuels obtained have a range of heating values from 18.27 to 24.913 MJ/kg [70–73]. SCG calorific values are higher than other types of biomass, and it could be considered a viable fuel to cover the needs of thermal energy of the coffee industry [72].

6.2.2 Biodiesel production

Today, the world needs to change the fossil fuel dependence to renewable energy, as it is the case for biodiesel, which has less hydrocarbon, CO_2, and particle emission than conventional diesel [61]. New bioresources for biodiesel production are being explored, and SCG can be a viable alternative due to its high lipid containing 6–27.8% w/wt [61, 74, 75]. The biodiesel can be produced by transesterification of lipid and oil extracts. It is important to point out that biodiesel yield could be improved when catalysts and ultrasound-assisted processes are employed. The range of biodiesel yield obtained in different studies varied from 16.73 to 100% [63, 76, 77].

6.2.3 Bio-oil production

The main goal of this process is converting SCG into bio-oil. Fast pyrolysis, hydrothermal liquefaction in hot-compressed water, and co-liquefaction in sub-critical water have been tested. It is important to point out that in the pyrolysis process, bio-oil, water, biochar, and syngas are produced. The bio-oil yield obtained for these methods varied from 36 to 61.8 wt% of bio-oil [60, 69, 78–83]. The fast pyrolysis has been the method with the best bio-oil yield.

6.2.4 Biosyngas production

The SCG could be used to generate power and heat. This process is well known as cogeneration or combined heat and power process. It is used to satisfy the energy needs of industrial plants. The energy and power are generated by SCG gasification at moderate pressure (0.3–0.5 bar), temperature above 650°C, and using oxidants such as air, steam, and carbon dioxide. The gas produced of this reaction is named syngas or producer biogas, which contains methane, carbon dioxide, carbon monoxide, and hydrogen. The syngas can be burnt in a fuel cell or a conventional combustion engine [61, 84].

6.3 Biorefinery

Biomass revalorization via the conversion into value-added products and fuel is the main goal of a biorefinery, which is considered a sustainable process. The productivity maximization of intermediates and products is reached when an optimal sequence of multifunctional processes is integrated into the biorefinery. Then, the economics of waste revalorization is enhanced. The biorefinery uses several techniques and treatment methods for biomass conversion such as fermentation, extraction, hydrolysis, transesterification, and pyrolysis. It is important to point out that biological processes could also be used (fermentation, anaerobic digestion, etc.). A biorefinery could use the separation processes and unit operations of a petrochemical complex [3]. However, a biorefinery is highly dependent on biomass composition, availability, and the economic value of bioproducts obtained [3, 60, 69].

A biorefinery could be an efficient method for obtaining valuable products from SCG due to its elemental composition; chemical composition (oil content, fatty acid, carbohydrates, carbonaceous and nitrogen compounds, etc.); low cost; high availability; and calorific value. The SCG could produce several value-added products (biosorbent, green composite, antioxidants, polyols, carotenoids, polyphenols, polyhydroxyalkanoates, polyurethane foam, Chlorogenic acid, tannins, activated carbon, PHA, caffeine, etc.) and bioenergy (biogas, biodiesel, and bio-oil). Attabani et al. proposed a biorefinery process using SCG as feedstock for obtaining biofuel, bioethanol, biogas, bio-oil, H_2, biodiesel, fuel pellets, biochar, polymers, compost, adsorbent, bioactive compounds, and pharmaceutical products [72]. The production of xylitol, activated carbon, phenolic acid, lactic acid, and heat using brewer's spent grain as the feedstock of a biorefinery [60, 85]. The process sustainability of biorefinery was demonstrated, thanks to the economic margin (62.25%), the potential environmental impact (0.012 PEI/kg products), and the carbon footprint (0.96 kg CO_2-e/kg of BSG).

7. Experimental design to process optimization

Response surface methodology (RSM) is a methodology used to improve process via very few essays, reducing cost and time. The RSM uses statistical and experimental design tools to obtain an optimal response, which is useful for making the right decision. The process performance is very complex due to numerous parameters that affect their behavior. RSM allows built process behavior maps based on mathematical models containing the significant parameters to achieve the maximum, target or minimum process performance.

The optimization of complex processes locates the best experimental conditions at which the process presents the minimum or maximum performance (yield, efficiency, etc.). The use of experimental design for optimizing processes has several advantages: less treatment time, low cost, and efficient use of resources, such as materials, equipment, and workforce. Besides, it uses tools of numerical regression to fit the data to mathematical models to predict values on the region of studied factor levels.

7.1 Use of experimental design on coffee waste

The experimental design has been used to optimize the extraction conditions of coffee parchment waste (CP) [86], antioxidant phenolic compounds from coffee silverskin (CS) [8], total phenolic compound and caffeine from SCG [86], coffee oil from SCG [87], the removal conditions of free fatty acid of SCG [88], the conditions to reducing sugar from SCG [89], organic acids [90] and alcohol production from coffee waste [91], and the conditions for the quantification of heavy metals (Cd(II) and Pb(II)), where a carbon-paste electrode modified with SCG was used as a working electrode [92].

7.1.1 Type of experimental designs

The experimental design tools most used are the central composite design, the Box-Behnken design, and the Plackett-Burman design.

Box-Behnken experimental design was used by Mirón-Mérida et al. [86] to maximize the extract yield, total phenolic content, antioxidant activity, and caffeine content on CP simultaneously. The effects of three parameters on the responses were studied: liquid/solid ratio (10, 30, and 50), extraction temperature (45, 60, and 75°C), and ethanol percentage (50, 75, and 100%). The maximum extract yield of 2.36% was achieved at 75°C with 66.76% ethanol as a solvent and with 50 of liquid/solid ratio. The maximum caffeine extracted was 1.513 g caffeine kg^{-1} CP at 74.35°C and 69.64% ethanol with 33.47 of liquid/solid ratio. The highest total phenolic content of 2.84037 g gallic acid kg^{-1} CP was obtained at 14.33 liquid/solid ratio, 70.74% ethanol, and 75°C. For the maximum extraction of 12.69 μmol Trolox g^{-1}, CP of antioxidant activity was attained at liquid/solid ratio of 50, temperature of 75°C, and ethanol of 59.47%. Finally, the optimal extraction conditions were established at 75°C with 41 liquid/solid ratio using 70% of aqueous ethanol as solvent.

Ballesteros et al. [8] used a 23 face-centered central composite design to maximize the extraction of antioxidant phenolic compounds and oxidant activity from CS. The effects of ethanol concentration (20 and 90%), solvent/solid ratio (10 and 40 ml/g), and extraction time (90 and 30 min) were studied on the two responses. The highest phenolic compounds of 13 mg gallic acid equivalents/g CS, with the maximum antioxidant activity of 18.24 μmol Trolox equivalents/g CS and 0.83 mmol Fe(II)/g CS, were achieved at 60% ethanol as solvent, a ratio of 35 ml/g CS dry matter, during 30 min at 60–65°C [8].

Shang et al. [87] developed a two-stage experimental statistical analysis to optimize extraction conditions for total phenolics (mg/g) and caffeine (mg/g) from SCG. First, the process parameter was screened through a Plackett-Burman experiment design to identify the significant parameters of the pressurized liquid extraction method that affect the extraction efficiency, using six parameters at two levels: temperature (80 and 160°C), the concentration of ethanol in water (25 and 75%), extraction time (5 and 20 min), pressure (500 and 2500 psi), sample loading weight (0.5 and 2.5 g), and flush (20 and 100%). The most critical parameters affecting total phenolics and caffeine extraction were temperature and sample loading weight, at 95°C and 0.8 g, respectively. In the second optimization stage, a second-order central composite experimental design, employing the two significant parameters, was used to maximize the total phenolics and caffeine. The highest total phenolic compounds of 22.91 mg/g and caffeine extraction of 9.66 mg/g were achieved with 0.8 g sample loading weight at 195°C.

Pichai and Krit [88] applied response surface methodology to optimize the effects on the coffee oil yield for the solvent extraction process of the ratio of DSCG-hexane (1:8–1:22 g/g) and extraction time (6–34 min). According to the optimal conditions of 1:22.5 g/g mass ratio of DSCG-to-hexane and 30.4 min of extraction time under the 30°C of room temperature, the highest coffee oil yield estimated (14.75 wt%) and experimental (14.68 wt%) was reached.

Mueanmas et al. [89] used a central composite design to investigate the effect on the FFA removal percentage of the mole ratio (5–15) of MeOH-free fatty acid (FFA), the quantity of catalyst (5–15 wt%), the reaction temperature (50–70°C), and the reaction time (30–120 min). The maximum predicted (95.06%) and experimental (93.88%) of FFA removal was attained at 9.1:1 mol ratio of MeOH/FFA with 11.7 wt% of catalyst and 97.2 min of reaction time at 65°C.

Ravindran et al. [90] proposed a central composite design to maximize the reducing sugar yield of SCG, after enzymatic saccharification of pretreated biomass and ultrasound-assisted potassium permanganate oxidation. The effects of five parameters on the responses were studied: 77.08 FPU/mL of cellulase (biomass loading 1–5 g/50 ml), 72.23 U/mL of hemicellulase (biomass loading 0.3–1.5 ml/50 ml), pH (4.8–6.6), and incubation time (24–120 h). A maximum reducing sugar yield of 35.64 mg/mL of reaction volume was estimated with a high biomass loading of 5 g/50 mL, 1.5 mL/50 mL of cellulase, 0.37 mL/50 mL of hemicellulase, pH 6.7, and a low incubation time of 24 h. The experimental values obtained using the optimized parameters are in the range of total reducing sugar of 35.15 ± 0.2 mg/mL.

Montoya et al. [91] developed a Plackett-Burman design to evaluate the effect of the parameters on H_2, organic acids, and alcohol production from coffee waste. The coffee waste was pretreated using a consortium of bacteria and fungi (indigenous from coffee waste) with hydrolytic and fermentation activity in a hydrothermal reactor. The parameters of pH (4.0–7.0), temperature (30–50°C), agitation (0–180 rpm), headspace (50–70%), percentage of bioaugmentation (without microbial consortium to 20%), the concentration of coffee pulp and husk (2–6 g/L), coffee processing wastewater (7–30 g COD/L), and yeast extract (0–2 g/L) were studied. Under the optimum conditions of 30°C, 180 rpm, 50% headspace, without bioaugmentation, 2 g/L pulp and husk coffee, 30 gCOD/L coffee processing wastewater, and 2 g/L yeast extract, estimated production of 82 ml H_2 was achieved.

Finally, Estrada-Aldrete et al. [92] applied a central composite design to optimize the quantification of two heavy metals (Cd(II) and Pb(II)) at trace levels using a paste carbon electrode of spent coffee grounds, which was chemically modified by citric acid. The metal quantification was carried out by differential pulse anodic stripping voltammetry technique. The electrodeposition potential (−1200, −950, and −700 mV) and accumulation time (30, 75, and 120 s) were employed as design parameters. The optimal conditions to achieve the maximum Pb(II) anodic peak current of 2.09×10^{-4} A were − 1200 mV electrodeposition potential and 120 s accumulation time. The maximum Cd(II) anodic peak current of 1.385×10^{-3} A obtained at −1155 mV potential and 76 s time.

8. Conclusion

Coffee waste is widely available, and while it is being disposed of as domestic or industrial garbage, it represents a vital source to obtain valuable products and energy. Physico-chemical properties of coffee waste allow their revalorization in various applications, highlighting as a feedstock of biorefinery, due to the presence of useful chemical compounds; as a raw material in the synthesis of activated carbon, given the predominance of carbon; or applied directly as a biosorbent in pollutant removal from gas or liquid, thanks to its surface characteristics. The implementation of environmentally friendly processes based on coffee waste requires a deepening knowledge of the physico-chemical properties.

Coffee wastes are low-cost adsorbents for the removal of organic and inorganic pollutants from aqueous solutions in batch systems. However, more studies are needed to fully characterize the performance of coffee waste in continuous systems as fixed-bed columns to scale-up the process. Since coffee waste was found to be efficient in the removal of ozone, it is expected that future studies will focus on the application of coffee wastes in the removal of gaseous pollutants.

SCG activated carbon could be used in the adsorption process for removing organic and inorganic pollutants from aqueous solutions. According to recent literature analyzed, the activated carbon or biochar obtained from SCG shows excellent properties to be used as adsorbent materials, such as high surface area, wide pore, and total pore volume. Most of researchers have used an electric furnace to perform the carbonization process, which requires high power consumption; this represents an environmental liability because this production process leads to air pollution by greenhouse gases. Thus, it is necessary to increase the studies of the use of microwaves in the carbonization process. This technology requires a low time to perform the carbonization. Therefore, a low power consumption is needed.

An experimental design is a powerful tool to optimize systems where the mathematical relationships between the parameters and the process performance are unknown. Some attempts have been made to use them on the processing of coffee. However, it is necessary to use them to obtain optimal conditions for the recovery of valuable compounds on mono-process extraction before the implementation of a biorefinery.

Experimental design methodology could help to obtain a sustainable process not only in the revalorization of coffee waste but also in all the stages of coffee processing.

Author details

Felipe J. Cerino-Córdova[1*], Nancy E. Dávila-Guzmán[2], Azucena M. García León[2], Jacob J. Salazar-Rabago[2] and Eduardo Soto-Regalado[2]

1 Facultad de Ingeniería Mecánica y Eléctrica (FIME), CIDIIT, Universidad Autónoma de Nuevo León (UANL), San Nicolás de Los Garza, Nuevo León, Mexico

2 Facultad de Ciencias Químicas (FCQ), Universidad Autónoma de Nuevo León (UANL), San Nicolás de Los Garza, Nuevo León, Mexico

*Address all correspondence to: felipe.cerinocr@uanl.edu.mx; felipejccuanl@yahoo.com.mx

References

[1] Levy D, Reinecke J, Manning S. The political dynamics of sustainable coffee: Contested value regimes and the transformation of sustainability. Journal of Management Studies. 2016;**53**: 364-401. DOI: 10.1111/joms.12144

[2] International Coffee Organization. http://www.ico.org/ [Accessed: 02 August 2019]

[3] Mata TM, Martins AA, Caetano NS. Bio-refinery approach for spent coffee grounds valorization. Bioresource Technology. 2018;**247**:1077-1084. DOI: 10.1016/j.biortech.2017.09.106

[4] Murthy PS, Naidu MM. Sustainable management of coffee industry by-products and value addition—A review. Resources, Conservation and Recycling. 2012;**66**:45-58. DOI: 10.1016/j.resconrec.2012.06.005

[5] Pujol D, Liu C, Gominho J, Olivella MA, Fiol N, Villaescusa I, et al. The chemical composition of exhausted coffee waste. Industrial Crops and Products. 2013;**50**:423-429. DOI: 10.1016/j.indcrop.2013.07.056

[6] Davila-Guzman NE, Cerino-Cordova FJ, Soto-Regalado E, Rangel-Mendez JR, Diaz-Flores PE, Garza-gonzález MT, et al. Copper biosorption by spent coffee ground: Equilibrium, kinetics, and mechanism. Clean: Soil, Air, Water. 2013;**41**(6): 557-564. DOI: 10.1002/clen.201200109

[7] Mussatto SI, Machado EMS, Martins S, Teixeira JA. Production, composition, and application of coffee and its industrial residues. Food and Bioprocess Technology. 2011;**4**:661-672. DOI: 10.1007/s11947-011-0565-z

[8] Ballesteros LF, Teixeira JA, Mussatto SI. Chemical, functional, and structural properties of spent coffee grounds and coffee silverskin. Food and Bioprocess Technology. 2014;7:3493-3503. DOI: 10.1007/s11947-014-1349-z

[9] Botello-González J, Cerino-Cordova FJ, Davila-Guzman NE, Salazar-Rabago JJ, Soto-Regalado E, Gomez-Gonzalez R, et al. Water, Air, and Soil Pollution. 2019;**230**:73. DOI: 10.1007/s11270-019-4106-0

[10] Alexander M. Introduction of Soil Microbiology. New York: John Wiley & Sons Inc.; 1961. pp. 472

[11] Volesky B. Biosorption and me. Water Research. 2007;**41**:4017-4029. DOI: 10.1016/j.watres.2007.05.062

[12] Ahsan A, Islam T, Imam MA, Hyder G, Jabbari V, Dominguez N, et al. Biosorption of bisphenol A and sulfamethoxazole from water using sulfonated coffee waste: Isotherm, kinetic and thermodynamic studies. Journal of Environmental Chemical Engineering. 2018;**6**:6602-6611. DOI: 10.1016/j.jece.2018.10.004

[13] Gomez-Gonzalez R, Cerino-Cordova FJ, Garcia-Leon AM, Soto-Regalado E, Davila-Guzman NE, Salazar-Rábago JJ. Lead biosorption onto coffee grounds: Comparative analysis of several optimization techniques using equilibrium adsorption models and ANN. Journal of the Taiwan Institute of Chemical Engineers. 2016;**68**:201-210. DOI: 10.1016/j.jtice.2016.08.038

[14] Matos AT, Brandão VS, Martinez MA, Neves JCL. Removal of Cu and Zn from swine raising wastewater using organic filters. Environonmental Technology (United Kingdom). 2003;**24**:171-178. DOI: 10.1080/09593330309385548

[15] Tokimoto T, Kawasaki N, Nakamura T, Akutagawa J, Tanada S. Removal of lead ions in drinking water by coffee grounds as vegetable biomass. Journal of Colloid and Interface

Science. 2005;**281**:56-61. DOI: 10.1016/j.jcis.2004.08.083

[16] Oliveira WE, Franca AS, Oliveira LS, Rocha SD. Untreated coffee husks as biosorbents for the removal of heavy metals from aqueous solutions. Journal of Hazardous Materials. 2008;**152**:1073-1081. DOI: 10.1016/j.jhazmat.2007.07.085

[17] Malara A, Paone E, Frontera P, Bonaccorsi L, Panzera G, Mauriello F. Sustainable exploitation of coffee silverskin in water remediation. Sustainability. 2018;**10**. DOI: 10.3390/su10103547

[18] Azouaou N, Sadaoui Z, Djaafri A, Mokaddem H. Adsorption of cadmium from aqueous solution onto untreated coffee grounds: Equilibrium, kinetics and thermodynamics. Journal of Hazardous Materials. 2010;**184**:126-134. DOI: 10.1016/j.jhazmat.2010.08.014

[19] Elsherif KM, Haider I, El-Hashani A. Adsorption of Co(II) ions from aqueous solution onto tea and coffee powder: Equilibrium and kinectic studies. Journal of Fundamental Applied Sciences. 2019. DOI: 10.4314/jfas.v11i1.5

[20] Futalan CM, Kim J, Yee JJ. Adsorptive treatment via simultaneous removal of copper, lead and zinc from soil washing wastewater using spent coffee grounds. Water Science and Technology. 2019;**79**:1029-1041. DOI: 10.2166/wst.2019.087

[21] Mora Alvarez NM, Pastrana JM, Lagos Y, Lozada JJ. Evaluation of mercury (Hg^{2+}) adsorption capacity using exhausted coffee waste. Sustainable Chemistry and Pharmacy. 2018;**10**:60-70. DOI: 10.1016/j.scp.2018.09.004

[22] Kyzas GZ. Commercial coffee wastes as materials for adsorption of heavy metals from aqueous solutions. Materials (Basel). 2012;**5**:1826-1840. DOI: 10.3390/ma5101826

[23] Davila-Guzman NE, Cerino-Córdova FJ, Loredo-Cancino M, Rangel-Mendez JR, Gómez-González R, Soto-Regalado E. Studies of adsorption of heavy metals onto spent coffee ground: Equilibrium, regeneration, and dynamic performance in a fixed-bed column. International Journal of Chemical Engineering. 2016;**2016**:1-11. DOI: 10.1155/2016/9413879

[24] Oliveira LS, Franca AS, Alves TM, Rocha SDF. Evaluation of untreated coffee husks as potential biosorbents for treatment of dye contaminated waters. Journal of Hazardous Materials. 2008;**155**:507-512. DOI: 10.1016/j.jhazmat.2007.11.093

[25] Dai Y, Zhang K, Meng X, Li J, Guan X, Sun Q. New use for spent coffee ground as an adsorbent for tetracycline removal in water. Chemosphere. 2019;**215**:163-172. DOI: 10.1016/j.chemosphere.2018.09.150

[26] Utomo HD, Hunter KA. Adsorption of heavy metals by exhausted coffee grounds as a potential treatment method for waste waters. e-Journal of Surface Science and Nanotechnology. 2006;**4**:504-506. DOI: 10.1380/ejssnt.2006.504

[27] Hsieh PF, Wen TY. Evaluation of ozone removal by spent coffee grounds. Scientific Reports. 2020;**10**:1-8. DOI: 10.1038/s41598-019-56668-5

[28] Lafi R, Hafiane A. Removal of methyl orange (MO) from aqueous solution using cationic surfactants modified coffee waste (MCWs). Journal of the Taiwan Institute of Chemical Engineers. 2016;**58**:424-433. DOI: 10.1016/j.jtice.2015.06.035

[29] Cerino-Córdova FJ, Díaz-Flores PE, García-Reyes RB, Soto-Regalado E, Gómez-González R, Garza-González MT. Biosorption of Cu(II) and Pb(II) from aqueous solutions by chemically modified spent coffee grains. International

journal of Environmental Science and Technology. 2013;**10**:611-622. DOI: 10.1007/s13762-013-0198-z

[30] Naga BA, Reddy DS, Kumar GS, Ravindhranath K, Krishna Mohan GV. Removal of lead and fluoride from contaminated water using exhausted coffee grounds based bio-sorbent. Journal of Environmental Management. 2018;**218**:602-612. DOI: 10.1016/j.jenvman.2018.04.091

[31] Ahsan MA, Islam MT, Imam MA, Hyder AHMG, Jabbari V, Dominguez N. Biosorption of bisphenol A and sulfamethoxazole from water using sulfonated coffee waste: Isotherm, kinetic and thermodynamic studies. Journal of Environmental Chemical Engineering. 2018;**6**:6602-6611. DOI: 10.1016/j.jece.2018.10.004

[32] Demir DA, Gülçiçek O, Gören N. Optimization of adsorption for the removal of cadmium from aqueous solution using Turkish coffee grounds. International Journal of Environmental Research. 2019;**13**:861-878. DOI: 10.1007/s41742-019-00224-6

[33] Torres-Caban R, Vega-Olivencia CA, Alamo-Nole L, Morales-Irizarry D, Roman-Velazquez F, Mina-Camilde N. Removal of copper from water by adsorption with Calcium-Alginate/Spent-Coffee-Grounds composite beads. Materials (Basel). 2019;**12**. DOI: 10.3390/ma12030395

[34] Torres-Caban R, Vega-Olivencia CA, Mina-Camilde N. Adsorption of Ni^{2+} and Cd^{2+} from water by calcium alginate/spent coffee grounds composite beads. Applied Sciences. 2019;**9**. DOI: 10.3390/app9214531

[35] Lessa EF, Nunes ML, Fajardo AR. Chitosan/waste coffee-grounds composite: An efficient and eco-friendly adsorbent for removal of pharmaceutical contaminants from water. Carbohydrate Polymers. 2018;**189**:257-266. DOI: 10.1016/j.carbpol.2018.02.018

[36] Le VT, Pham TM, Doan VD, Lebedeva OE, Nguyen HT. Removal of Pb(II) ions from aqueous solution using a novel composite adsorbent of Fe_3O_4/PVA/spent coffee grounds. Separation Science and Technology. 2019;**54**:3070-3081. DOI: 10.1080/01496395.2019.1565770

[37] Edathil AA, Shittu I, Hisham ZJ, Banat F, Haija MA. Novel magnetic coffee waste nanocomposite as effective bioadsorbent for Pb(II) removal from aqueous solutions. Journal of Environmental Chemical Engineering. 2018;**6**:2390-2400. DOI: 10.1016/j.jece.2018.03.041

[38] Iakovleva E, Sillanpää M, Maydannik P, Liu JT, Allen S, Albadarin AB. Manufacturing of novel low-cost adsorbent: Co-granulation of limestone and coffee waste. Journal of Environmental Management. 2017;**203**:853-860. DOI: 10.1016/j.jenvman.2017.05.039

[39] Liu L, Yu L, Zhang W, Fan J, Zuo Q, Li M. Adsorption performance of Pb(II) ions from aqueous solution onto a novel complex of coffee grounds and attapulgite clay. Desalination and Water Treatment. 2019;**153**:208-215. DOI: 10.5004/dwt.2019.24080

[40] Kawamoto H. Lignin pyrolysis reactions. Journal of Wood Science. 2017;**63**:117-132. DOI: 10.1007/s10086-016-1606-z

[41] Wang S, Ru B, Lin H, Luo Z. Degradation mechanism of monosaccharides and xylan under pyrolytic conditions with theoretic modeling on the energy profiles. Bioresource Technology. 2013;**143**:378-383

[42] Lin Y-C, Cho J, Tompsett GA, Westmorel PR, Huber GW. Kinetics

and mechanism of cellulosepyrolysis. The Journal of Physical Chemistry. 2009;**113**:20097-20107

[43] Gratuito MKB, Panyathanmaporn, Chumnanklang RA, Sirinuntawittaya N, Dutta A. Production of activated carbon from coconut shell: Optimization using response surface methodology. Bioresource Technology. 2011;**99**(11): 5887-4895

[44] Shuxiong T, Yao C, Xie R, Wenju J, Yanxin J. Preparation of activated carbon from corn cob and its adsorption behavior on Cr(VI) removal. Water Science and Technology. 2016;**73**(11):2654-2661. DOI: 10.2166/wst.2016.120

[45] Silva LV, López SJB, Correa-Guimaraes A, Hernández-Navarro S, Sánchez-Báscones M, Navas-García LM, et al. Rhodamine B removal with activated carbons obtained from lignocellulosic waste. Journal of Environmental Management. 2015;**155**:67-76. DOI: 10.1016/j.jenvman.2015.03.007

[46] Fotouhi TN, Javier AS, Yohannes K. Coffee extract residue for production of ethanol and activated carbons. Journal of Cleaner Production. 2015;**91**:64-70. DOI: 10.1016/j.jclepro.2014.12.031

[47] Arunina N, Brij B, Vartika G, Sharma P. Chemically activated carbon from lignocellusic wastes for heavy metal wastewater remediation: Effect of activation conditions. Journal of Colloid and Interface Science. 2017;**493**:228-240. DOI: 10.1016/j.jcis.2017.01.031

[48] Kemp CK, Baek BS, Lee WG, Meyyappan M, Kwang KS. Activated carbon derived from waste coffee grounds for stable methane storage. Nanotechnology. 2015;**26**(38):1-12. DOI: 10.11088/0957-4485/26/38/385602

[49] Fonseca A, Andreira C, Pacheco A, Romario V, de Oliveria SB, Ojala AS, et al. Activated carbon produce from waste coffee grounds for an effective removal of bispheno-A in aqueouos solutions. Environmental Science and Pollution Research. 2019;**26**(23):24850-24862

[50] Díaz de León ML, Soto-Regalado E, Loredo-Cancino M, Cerino-Córdova FJ, García Reyes RB, Dávila Guzmán NE, et al. Phenol adsorption onto coffee waste-granular activated carbon: Kinetics and equilibrium studies in aqueous solutions. Desalination and Water Treatment. 2017;**90**:231-240. DOI: 10.5004/dwt.2017.21418

[51] Jutakridsada P, Prajaksud C, Kuboonya-Aruk L, Theerakulpisut KK. Adsorption characteristics of activated carbon prepared from spent ground coffee. Clean Technologies and Environmental Policy. 2016;**18**(3):639-645. DOI: 10.1007/s10098-015-1083-x

[52] Kante C, Nieto-Delgado C, Rangel-Mendez JR, Bandosz TJ. Spent coffee-based activated carbon: Specific surface features and their importance for H_2S separation process. Journal of Hazardous Materials. 2012;**201**:141-147. DOI: 10.1016/j.jhazmat.2011.11.053

[53] Gadipelli TW, Zhengxiao G. Superior CO_2 adsorption from waste coffee ground derived carbons. RSC Advances. 2015;**5**(37):29558-29562. DOI: 10.1039/c4ra13026

[54] Laksaci H, Khelifi A, Trari M, Addoun A. Synthesis and characterization of microporous activated carbon from coffee grounds using potassium hydroxides. Journal of Cleaner Production. 2017;**147**(20):254-262. DOI: 10.1016/j.jclepro.2017.01.102

[55] Ching SL, Yusoff SM, Aziz AH, Umar M. Influence of impregnation ratio on coffee ground activated carbon as landfill leachate adsorbent for removal of total iron and of thophospafe. Desalination. 2011;**279**:225-234

[56] Plaza MG, González AS, Pevidia C, Pis JJ, Rubiera F. Valorisation of spent coffee grounds as CO_2 adsorbents for postcombustion capture applications. Applied Energy. 2012;**99**:272-279. DOI: 10.1016/j.apenergy.2012.05.028

[57] Boudrahem F, Soualah A, Aissani-Benissad F. Pb(II) and Cd(II) removal from aqueous solutions using activated carbon developed from coffee residue activated with phosphoric acid and zinc chloride. Journal of Chemical & Engineering Data. 2011;**56**:1946-1955. DOI: 10.1021/je1009569

[58] Cooney DO. Adsorption Design for Wastewater Treatment. CRC Press LLC; 1999. ISBN: 1-56670-333-6

[59] Philp J, Winickoff DE. Realising the circular bioeconomy. OECD Science, Technology and Industry Policy Papers. 2018;**60**:1-74. DOI: 10.1787/31bb2345-en

[60] Zabaniotou A, Kamaterou P. Food waste valorization advocating circular bioeconomy—A critical review of potentialities and perspectives of spent coffee grounds biorefinery. Journal of Cleaner Production. 2019;**211**:1553-1566

[61] Kourmentza C, Economou CN, Tsafrakidou P, Kornaros M. Spent coffee grounds make much more than waste: Exploring recent advances and future exploitation strategies for the valorization of an emerging food waste stream. Journal of Cleaner Production. 2018;**172**:980-992. DOI: 10.1016/j.jclepro.2017.10.088

[62] Campos-Vega R, Loarca-Piña G, Vergara-Castañeda H, Oomah BD. Spent coffee grounds: A review on current research and future prospects. Trends in Food Science and Technology. 2015;**45**(1):24-36. DOI: 10.1016/j.tifs.2015.04.012

[63] Ponte Rocha MV, Brandão Lima de Matos LJ, Pinto de Lima L, da Silva Figueiredo PM, Lucena IL, Narciso Fernandes FA, et al. Ultrasound-assisted production of biodiesel and ethanol from spent coffee grounds. Bioresource Technology. 2014;**167**:343-348. DOI: 10.1016/j.biortech.2014.06.032

[64] Azmir J, Zaidul ISM, Rahman MM, Sharif KM, Mohamed A, Sahena F, et al. Techniques for extraction of bioactive compounds from plant materials: A review. Journal of Food Engineering. 2013;**117**(4):426-436. DOI: 10.1016/j.jfoodeng.2013.01.014

[65] Georgieva SS, Coelho JAP, Campos FC, Robalo MP, Stateva RP. Green extraction of high added value substances from spent coffee grounds: Preliminary results. Journal of Chemical Technology and Metallurgy. 2018;**53**(4):640-646

[66] Massaya J, Pereira AP, Mills-Lamptey B, Benjamin J, Chuck CJ. Conceptualization of a spent coffee grounds biorefinery: A review of existing valorisation approaches. Food and Bioproducts Processing. 2019;**118**:149-166. DOI: 10.1016/j.fbp.2019.08.010

[67] Simões J, Nunes FM, Domingues MR, Coimbra MA. Extractability and structure of spent coffee ground polysaccharides by roasting pre-treatments. Carbohydrate Polymers. 2013;**97**(1):81-89. DOI: 10.1016/j.carbpol.2013.04.067

[68] Passos CP, Coimbra MA. Microwave superheated water extraction of polysaccharides from spent coffee grounds. Carbohydrate Polymers. 2013;**94**:626-633. DOI: 10.1016/j.carbpol.2013.01.088

[69] Rajesh BJ, Kavitha S, Yukesh KR, Dinesh KM, Preethi AAE, Kumar G. Biorefinery of spent coffee grounds waste: Viable pathway towards circular bioeconomy. Bioresource Technology. 2020;**302**:1-15. DOI: 10.1016/j.biortech.2020.122821

[70] Ciesielczuk T, Karwaczynska U, Sporek M. The possibility of disposing of spent coffee ground with energy recycling. Journal of Ecological Engineering. 2015;**16**:133-138. DOI: 10.12911/22998993/59361

[71] Jeguirim M, Bikai J, Elmay Y, Limousy L, Njeugna E. Thermal characterization and pyrolysis kinetics of tropical biomass feedstocks for energy recovery. Energy for Sustainable Development. 2014;**23**:188-193. DOI: 10.1016/j.esd.2014.09.009

[72] Atabani AE, Shobana S, Mohammed MN, Uğuz G, Kumar G, Arvindnarayan S, et al. Integrated valorization of waste cooking oil and spent coffee grounds for biodiesel production: Blending with higher alcohols, FT-IR, TGA, DSC and NMR characterizations. Fuel. 2019;**244**: 419-430. DOI: 10.1016/j.fuel.2019.01.169

[73] Limousy L, Jeguirim M, Dutournié P, Kraiem N, Lajili M, Said R. Gaseous products and particulate matter emissions of biomass residential boiler fired with spent coffee grounds pellets. Fuel. 2013;**107**:323-329. DOI: 10.1016/j.fuel.2012.10.019

[74] Efthymiopoulos I, Hellier P, Ladommatos N, Russo-Profili A, Eveleigh A, Aliev A, et al. Influence of solvent selection and extraction temperature on yield and composition of lipids extracted from spent coffee grounds. Industrial Crops and Products. 2018;**119**:49-56. DOI: 10.1016/j.indcrop.2018.04.008

[75] Kwon EE, Yi H, Jeon YJ. Sequential co-production of biodiesel and bioethanol with spent coffee grounds. Bioresource Technology. 2013;**136**:475-480. DOI: 10.1016/j.biortech.2013.03.052

[76] Caetano NS, Silva VFM, Mata TM. Valorization of coffee grounds for biodiesel production. Chemical Engineering Transactions. 2012:267-272. DOI: 10.3303/CET1226045

[77] Park J, Kim B, Lee JW. In-situ transesterification of wet spent coffee grounds for sustainable biodiesel production. Bioresource Technology. 2016;**221**:55-60. DOI: 10.1016/j.biortech.2016.09.001

[78] Kelkar S, Saffron CM, Chai L, Bovee J, Stuecken TR, Garedew M, et al. Pyrolysis of spent coffee grounds using a screw-conveyor reactor. Fuel Processing Technology. 2015;**137**:170-178

[79] Bok JP, Choi HS, Choi YS, Park HC, Kim SJ. Fast pyrolysis of coffee grounds: Characteristics of product yields and biocrude oil quality. Energy. 2012;**47**: 17-24

[80] Li X, Strezov V, Kan T. Energy recovery potential analysis of spent coffee grounds pyrolysis products. Journal of Analytical and Applied Pyrolysis. 2014;**110**:79-87

[81] Yang L, Nazari L, Yuan Z, Corscadden K, Xu CC. Hydrothermal liquefaction of spent coffee grounds in water medium for bio-oil production. Biomass and Bioenergy. 2016;**86**:191-198

[82] Yang L, He QS, Havard P, Corscadden K, Xu CC, Wang X. Co-liquefaction of spent coffee grounds and lignocellulosic feedstocks. Bioresource Technology. 2017;**237**:108-121. DOI: 10.1016/j.biortech.2017.02.087

[83] Cho DW, Tsang DCW, Kim S, Kwon EE, Kwon G, Song H. Thermochemical conversion of cobalt-loaded spent coffee grounds for production of energy resource and environmental catalyst. Bioresource Technology. 2018;**270**:346-351. DOI: 10.1016/j.biortech.2018.09.046

[84] Pacioni TR, Soares D, Di Domenico M, Rosa MF, De Fátima R, Muniz P, et al. Bio-syngas production

from agro-industrial biomass residues by steam gasification. Waste Management. 2016;**58**:221-229. DOI: 10.1016/j.wasman.2016.08.021

[85] García CA, Moncada J, Aristizábal V, Cardona CA. Techno-economic and energetic assessment of hydrogen production through gasification in the Colombian context: Coffee cut-stems case. International Journal of Hydrogen Energy. 2017;**42**:5849-5864. DOI: 10.1016/j.ijhydene.2017.01.073

[86] Mirón-Mérida VA, Yáñez-Fernández J, Montañez-Barragán B, Huerta BEB. Valorization of coffee parchment waste (*Coffea arabica*) as a source of caffeine and phenolic compounds in antifungal gellan gum films. LWT. 2019;**101**:167-174. DOI: 10.1016/j.lwt.2018.11.013

[87] Shang YF, Xu JL, Lee WJ, Um BH. Antioxidative polyphenolics obtained from spent coffee grounds by pressurized liquid extraction. South African Journal of Botany. 2017;**109**:75-80. DOI: 10.1016/j.sajb.2016.12.011

[88] Pichai E, Krit S. Optimization of solid-to-solvent ratio and time for oil extraction process from spent coffee grounds using response surface methodology. ARPN Journal of Engineering and Applied Sciences. 2015;**10**(16):7049-7052

[89] Mueanmas C, Nikhom R, Petchkaew A, Iewkittayakorn J, Prasertsit K. Extraction and esterification of waste coffee grounds oil as non-edible feedstock for biodiesel production. Renewable Energy. 2019;**133**:1414-1425. DOI: 10.1016/j.renene.2018.08.102

[90] Ravindran R, Jaiswal S, Abu-Ghannam N, Jaiswal AK. Evaluation of ultrasound assisted potassium permanganate pre-treatment of spent coffee waste. Bioresource Technology. 2017;**224**:680-687. DOI: 10.1016/j.biortech.2016.11.034

[91] Montoya ACV, da Silva Mazareli RC, Delforno TP, Centurion VB, Sakamoto IK, de Oliveira VM, et al. Hydrogen, alcohols and volatile fatty acids from the co-digestion of coffee waste (coffee pulp, husk, and processing wastewater) by applying autochthonous microorganisms. International Journal of Hydrogen Energy. 2019;**44**(39):21434-21450. DOI: 10.1016/j.ijhydene.2019.06.115

[92] Estrada-Aldrete J, Hérnández-López JM, García-León AM, Peralta-Hernández JM, Cerino-Córdova FJ. Electroanalytical determination of heavy metals in aqueous solutions by using a carbon paste electrode modified with spent coffee grounds. Journal of Electroanalytical Chemistry. 2020;**857**:1-10. DOI: 10.1016/j.jelechem.2019.113663